Artificial Intelligence and Knowledge Processing: Methods and Applications

Edited by

Hemachandran K.
Department of Artificial Intelligence
School of Business
Woxsen University
Hyderabad
India

Raul V. Rodriguez
School of Business
Woxsen University
Hyderabad
India

Umashankar Subramaniam
College of Engineering
Prince Sultan University
Riyadh
Saudi Arabia

&

Valentina Emilia Balas
Aurel Vlaicu University of Arad,
Arad,
Romania

Artificial Intelligence and Knowledge Processing: Methods and Applications

Editors: Hemachandran K., Raul V. Rodriguez, Umashankar Subramaniam & Valentina Emilia Balas

ISBN (Online): 978-981-5165-73-9

ISBN (Print): 978-981-5165-74-6

ISBN (Paperback): 978-981-5165-75-3

© 2023, Bentham Books imprint.

Published by Bentham Science Publishers Pte. Ltd. Singapore. All Rights Reserved.

First published in 2023.

need for a court order if at any point you breach any terms of this License Agreement. In no event will any delay or failure by Bentham Science Publishers in enforcing your compliance with this License Agreement constitute a waiver of any of its rights.

3. You acknowledge that you have read this License Agreement, and agree to be bound by its terms and conditions. To the extent that any other terms and conditions presented on any website of Bentham Science Publishers conflict with, or are inconsistent with, the terms and conditions set out in this License Agreement, you acknowledge that the terms and conditions set out in this License Agreement shall prevail.

Bentham Science Publishers Pte. Ltd.
80 Robinson Road #02-00
Singapore 068898
Singapore
Email: subscriptions@benthamscience.net

BENTHAM
SCIENCE

CONTENTS

FOREWORD

Artificial Intelligence & Knowledge Processing is playing a vital role in changing most of the sectors' processes and landscapes. AI has an enormous impact on various automation industries and their functioning converting traditional industries to AI-based factories. New algorithms are changing the way business processes and results are analyzed. You will encounter a few of the topics such as AI in Robotics, AI in IoT, AI in Marketing and Operations, AI in Healthcare, how insights are extracted from bigdata, Museums including Lighting and Cooling Systems, forecast visitor behavior, management of security systems and economic use of energy and other resources, automated path planning in a garden environment using reinforcement learning and analysis of human gait by utilizing regression in this collection of articles and papers. You will undoubtedly like them as much as I did when taking part, I'm sure. With its numerous research articles and chapters on Artificial Intelligence and Knowledge Processing: Methods and Applications, which is applicable to and helpful in our current world, I hope this book makes a significant contribution to various sectors.

Somdip Dey
Nosh Technologies
Colchester, United Kingdom

PREFACE

Artificial Intelligence has been gradually invading our lives, and we are using it to accomplish many fundamental tasks. Talking to Siri or Alexa, and avoiding traffic using Google Maps are the most common examples of Artificial Intelligence in our daily lives.

Intelligence, as a quality possessed by humans, is the ability to acquire and apply knowledge. This knowledge is gained through observation, perception, application of reasoning and logic to the observed facts. Artificial intelligence transforms how knowledge is captured, developed, shared, and efficiently used within organisations.

Many tools are used in AI, including mathematical optimization, logic, and probability and dwell in the fields of computer science, psychology, linguistics and many other subjects.

The major application of AI in Robotics, IoT, Marketing and Operations, how insights are extracted from bigdata, Museums including Lighting and Cooling Systems, forecast visitor behavior, management of security systems and economic use of energy and other resources, automated path planning in a garden environment using reinforcement learning and analysis of human gait by utilizing regression are discussed in this book. This book also gives insights into violence detection using computer vision for smart cities.

AI is also increasingly used in the health sector. For example, Artificial Neural Network is used to predict the presence or absence of heart disease. Scientists have used AI to predict crop suitability and rainfall prediction, referred to as a Crop Management system. Crop Management System uses parameters like the ratio of nitrogen, phosphorous, potassium, pH value of soil, and environmental factors such as humidity, rainfall, temperature, *etc.* These issues are also discussed in this book.

This book written in simple language throws light on the AI applications in several fields, including machine learning, medicine, and agriculture and briefs how AI impacts daily life. We hope that this book will be beneficial for students, researchers and all people who are interested in Artificial Intelligence and Knowledge Processing. We express our sincere thanks to the editorial and production teams who worked relentlessly to publish this book on time.

Hemachandran K.
Department of Artificial Intelligence
School of Business
Woxsen University
Hyderabad
India

Raul V. Rodriguez
School of Business
Woxsen University
Hyderabad
India

Umashankar Subramaniam
College of Engineering
Prince Sultan University
Riyadh
Saudi Arabia

&

Valentina Emilia Balas
Aurel Vlaicu University of Arad,
Arad
Romania

List of Contributors

Abhishek Rajput	Vishwakarma Institute of Technology, Pune, Maharashtra, India
Aditya Rasal	Vishwakarma Institute of Technology, Pune, Maharashtra, India
Amrapali Nimsarkar	Department of Electrical Engineering, GHRCE, Nagpur, India
B. R. S. S. Sowjanya	Woxsen School of Business, Woxsen University, Kamkole, Sadasivpet, Telangana, India
C. Roshan Abraham	Madras Institute of Technology, Anna University, Chennai, India
Chithik Raja	University of Technology and Applied Sciences Salalah, Salalah, Sultanate of Oman
Chinna Swamy Dudekula	Engineering and Environment (Advanced Computer Science with Advanced Practice), Northumbria University, Newcastle, United Kingdom
Choppala Swathi Priya	Woxsen School of Business, Woxsen University, Kamkole, Sadasivpet, Telangana-500078, India
Dibya Nandan Mishra	Woxsen School of Business, Woxsen University, Kamkole, Sadasivpet, Telangana, India
Devendra Dheeraj Gupta Sanagapalli	Department of Physics, BITS Pilani Hyderabad Campus, Telangana-500078, India
Ezendu Ariwa	University of Wales Trinity Saint David, London Campus, United Kingdom
Gabriel Kabanda	School of Business, Woxsen University, Hyderabad, India
Guda Vineeth Reddy	School of Business, Woxsen University, Hyderabad, India
Gaddam Venkat Shobika	Woxsen School of Business, Woxsen University, Kamkole, Sadasivpet, Telangana, India
Hemachandran K.	Department of Artificial Intelligence, School of Business, Woxsen University, Hyderabad, India
Harikumar Naidu	Department of Electrical Engineering, GHRCE, Nagpur, India
Ishank Jha	Woxsen School of Business, Woxsen University, Kamkole, Sadasivpet, Telangana-500078, India
Jyoti Madake	Vishwakarma Institute of Technology, Pune, Maharashtra, India
K. Jarina Begum	Jazan University, Jazan, Kingdom of Saudi Arabia
Keerti Adapa	Faculty of Science and Technology, ICFAI Foundation for Higher Education, Hyderabad, India
K. Vishal	Woxsen School of Business, Woxsen University, Kamkole, Sadasivpet, Telangana, India
K. Ravi Kumar Reddy	Lifencrypt, Hyderabad, India
K. Kailash	Independent Journalist, Hyderabad, India
Mamta Tembhare	Waste Processing Division, CSIR-NEERI, Nagpur, India

Mukul Kumar Gupta	School of Engineering, University of Petroleum and Energy Studies (UPES), Dehradun, India
Nitesh Singh Malan	School of Engineering, University of Petroleum and Energy Studies (UPES), Dehradun, India
Pingili Sravya	School of Business, Woxsen University, Hyderabad, India
P.K. Thiruvikraman	Department of Physics, BITS Pilani Hyderabad Campus, Telangana-500078, India
Piyush Kokate	Energy & Resource management Division, CSIR-NEERI, Nagpur, India
Pranay Kumar	University of Maryland, Baltimore County, United States
Pranav Balakrishnan	Madras Institute of Technology, Anna University, Chennai, India
Prathyusha Pujari	Woxsen School of Business, Woxsen University, Kamkole, Sadasivpet, Telangana, India
Rajesh Kumar K. V.	School of Business, Woxsen University, Hyderabad, India
Simran Sahni	Department of Physics, BITS Pilani Hyderabad Campus, Telangana-500078, India
Shripad Bhatlawande	Vishwakarma Institute of Technology, Pune, Maharashtra, India
Sambodhi Umare	Vishwakarma Institute of Technology, Pune, Maharashtra, India
Sri Rama Sai Pavan Kumar	School of Business, Woxsen University, Hyderabad, India
Sai Santosh Potnuru	Woxsen School of Business, Woxsen University, Kamkole, Sadasivpet, Telangana-500078, India
Sailaja Maggidi	School of Business, Woxsen University, Hyderabad, India
Sourav Chakraborty	Woxsen School of Business, Woxsen University, Kamkole, Sadasivpet, Telangana, India
Swati Shilaskar	Vishwakarma Institute of Technology, Pune, Maharashtra, India
Sandip K. Chourasiya	University of Petroleum and Energy Studies, Dehradun, India
Sudheer Hanumanthakari	Faculty of Science and Technology, ICFAI Foundation for Higher Education, Hyderabad, India
S. Sathiya Murthi	Madras Institute of Technology, Anna University, Chennai, India
Thakur Santosh	Mahindra University, Hyderabad, India
V. Sathiesh Kumar	Madras Institute of Technology, Anna University, Chennai, India
Varun Shelke	Vishwakarma Institute of Technology, Pune, Maharashtra, India
V. Devarajan	University of Technology and Applied Sciences Salalah, Salalah, Sultanate of Oman
Varadharaja Krishna	Woxsen School of Business, Woxsen University, Kamkole, Sadasivpet, Telangana, India
Y. Vani	NLP at Google, California, USA

CHAPTER 1

Artificial General Intelligence; Pragmatism or an Antithesis?

K. Ravi Kumar Reddy[1,*], **K. Kailash**[2] and **Y. Vani**[3]

[1] *Lifencrypt, Hyderabad, India*

[2] *Independent Journalist, Hyderabad, India*

[3] *NLP at Google, California, USA*

Abstract: Artificial intelligence is promoted by means of incomprehensible advocacy through business majors that cannot easily be equated with human consciousness and abilities. Behavioral natural systems are quite different from language models and numeric inferences. This paper reviews through centuries of evolved human knowledge, and the resolutions as referred through the critics of mythology, literature, imagination of celluloid, and technical work products, which are against the intellect of both educative and fear mongering. Human metamorphic abilities are compared against the possible machine takeover and scope of envisaged arguments across both the worlds of 'Artificial Intelligence' and 'Artificial General Intelligence' with perpetual integrations through 'Deep Learning' and 'Machine Learning', which are early adaptive to 'Artificial Narrow Intelligence' — a cross examination of hypothetical paranoid that is gripping humanity in modern history. The potentiality of a highly sensitive humanoid and sanctification to complete consciousness at par may not be a near probability, but social engineering through the early stages in life may indoctrinate biological senses to a much lower level of ascendancy to Artificial Narrow Intelligence — with furtherance in swindling advancement in processes may reach to a pseudo-Artificial Intelligence {i}. There are no convincing answers to the discoveries from ancient scriptures about the consciousness of archetypal humans against an anticipated replication of a fulfilling Artificial Intelligence {ii}. Human use of lexicon has been the focal of automata for the past few years and the genesis for knowledge, and with the divergence of languages and dialects, scores of dictionaries and tools that perform bidirectional voice and text — contextual services are already influencing the lives, and appeasement to selective humanly incidentals is widely sustainable today {iii}. Synthesizing and harmonizing a pretentious labyrinthine gizmo is the center of human anxiety, but only evaluative research could corroborate that tantamount to genetic consciousness.

Keywords: Brain, Dabus, Darpa, Enzymatic distress, Singularity, Technophobia, Upanishads.

---

* Corresponding author K. Ravi Kumar Reddy: Lifencrypt, Hyderabad, India; E-mail: cs@lifencrypt.com

Hemachandran K., Raul V. Rodriguez, Umashankar Subramaniam & Valentina Emilia Balas (Eds.)

INTRODUCTION

Human, the conditioned being to near living of earthly matters, is expected to become a mere biological slave in front of the mammoth data insiders where machines to become brainy, and with very sturdy self-corrective aptitudes to work against the Darwinian doctrine of species. The hypothetical apocalypse of a kind that is expected to dominate all biological inhibitions and races to reduce to nothing but menials. So far, superior being on the surface of the earth system, homo-sapiens, appraised to get into deep descent in their metamorphic intelligence being transmuted to the ownership of machines. The egoistic behavior of creative gizmo with intelligence and superlative augmentation of digital proficiency is meant for reduction of civilization to an imminent atrophy. Forced authority with an invisible antagonist in the form of a burgeoning digital consciousness is foreseen to be a certainty against future genealogy. These are pragmatic proclamations, though — the reality could well be very much a disregard for the prospects of dream merchants.

Dutch artist Maurits C. Escher (1898-1972) is famous for his lithographic depiction of 'drawing hands' and is the greatest testimony to contemplation of exertion towards a lifeless scheme to be judicious. He was not of any great repute by the art world but was idolized by mathematicians [1]. The synopsis constitutes a perfect example of a genuinely strange loop, but this conviction arises from our suspended skepticism and psychic sneak into Escher's intriguing domain [2] (p.103). The venerated mastery in Escher's artistry is envisioned for perplexity across unattainability and pragmatism. Logico-semantic paradoxes may stretch a supposition to approval in the coincidence of a parallelly persistent unacceptability as contrary [3] (pp.299-308). This innate fortitude through ardent stimulus by frontal lobes for temporal paradox and reflexes is preserved to be a unique ability by humans as a biological competency that may be compared to none of artificiality and in generations to survive. Among the greatest of abilities, an unusual trait of a Liar Paradox[1] is very significant in reciprocal transactions among human intelligence — and for sure as polymorphism in egotistical motivations. The Liar had to be regarded of every circumstance as a statement that is both true and false [4], a true anomalous contradiction.

Every endeavor of the human is not definitely effortless since it warps through the multitude of fluidity and complexity of logical amplitude. The antiquity and chronicled experiences among races in each case of every being tend to be unique and unparallel. Indeed, the bizarre dilemma of annoyance, John Locke's[2] empirical epistemology symbolizes a logical and political conclusion, with advanced theories from Wollstonecraft[3] and Godwin[4] in which humans figured already as artificially altered creatures who deserve liberation from the tyranny of political systems that had perverted and contorted their biological intelligence [5] (p.183).

Artificial Life (A-Life) is a theoretical intelligence in observations of natural occurrences to biological life — explicit to the most apparent composition of living thing, and aptitude to self-regulate as an instantaneous buildup of form and structure [6] (p.507). Langton[5] articulates that A-Life is the study relating to the replication of natural living systems through man-made appliances and would complement the synthesized biological demeanor by artificial means. Contrary, Darwinian biological evolution has many pathways in determining species across millions of years, and at each juncture of transfiguration in relinquishing those unwanted and acquisition of required abilities for continuing genetic lineage — and to have been an exhaustive fulfilment of skill in ecological conditioning, which is seized to temporal stature with categorical constraints of survival.

But enthusiastic at the arrival of the computation machine, in February 1951, Turing[6] wrote to one of his colleagues at the National Physical Laboratory, *"Our new machine [the Ferranti Mark I] is to start arriving on Monday. I am hoping as one of the first jobs to do something about 'chemical embryology'. In particular I think one can account for the appearance of Fibonacci numbers in connection with fir-cones".* These were free interpretations from the first-generation computing probabilities, and seven decades of information history must explain these ambitions. He certainly had a question of *"Can machines think?"*, but believed it to be meaningless to deserve discussion at his time. Turing was very cautious in his argument and believed that *"at the end of the century using these words and general educated opinion will have altered so much that one will be able to speak of machines thinking without expecting to be contradicted"* [7]. By the late twentieth century, humans started defining the scope and variants in the hierarchy of artificial to deliver intelligence, so ancient sages had their own definitions of intelligence and super intelligence in the holy scriptures.

ELUCIDATION OF ECHELONS, THE HIERARCHY

AnI is a very feeble AI and delivered in many forms, *i.e.,* Amazon, Netflix, Siri and Alexa are examples of task affirmative request comprehension. Smartphones, IoT, and pet robots are also some other forms of machines that may also be called AnI, since equipped with human supportive applications within their systems.

AI is implemented in machines to perform tasks that would require human intelligence to substantiate the functioning due to dynamics in presence, reasoning, likes and dislikes, learning, problem solving and quick decision making. AI is nothing but algorithms with certain sets of rules that try to learn the scope from the iteration of maturity in confidence level where the digital data learning (aka machine learning algorithms) is predetermined to the system.

AgI is a level of intelligence like the above ecological ability of human beings. It is to process and analyze knowledge to form psyche and reactions that can create irrevocable changes in humanity. AgI would be self-powered to access an AI digital brain connected to the internet and completely control humanity in imperceptible ways.

AsI, when AGI could accomplish complete control over humanity and gradually obtain a level of superintelligence, it can be so independent with extraordinary abilities and do not require any green thumb for catastrophic growth. AsI can perform unbelievable, according to scientists and come to have in bending atoms, destroy our world, rebuild it, and move and build machines or robotics at extremely fast rates while connected in real-time with other super-intelligent peers.

(Free adaptation from Cyrus A. Parsa, CEO, The AI Organization; and judiciously ameliorated classifications)

ANCIENT MYTHOLOGICAL MILIEU

Genesis 2:7; "The LORD God formed the man from the dust of the ground and breathed into his nostrils the breath of life, and the man became a living being." [8]. A sequence after the humans in theodicy of a pandora, the evil of AI — from the dawning of humans by God on this microcosm, man to devise machines, machines reinvent and refurbish to intelligent machines, mechanization to craft clayed inanimate to the life of superhumans, and stellar humans to... an insurmountable god matter with the phenomenon of a Prometheus[7], extravagantly a parabolic order. Since the ages of civilization, humans have been riding with the tide of mystical conception in replication to rebuild themselves, and automatons that mimic human faculties or alleviate them, or a desire to overpower evolution into own hands, or even play God. Dialectical Materialism[8] may refute any such fortuity since emphasizing the importance of earthly circumstances and the presence of contradictions within things. Again, idealist Hegelian Dialectic[9] emphasizes the observation that contradicts material singularities which could be resolved by juxtaposition and synthesis of a solution whilst retaining their essence, a probable superhuman/posthuman phenomenon.

The worship of the elephant-faced Ganesa, the son of Lord Siva and Goddess Parvati and the Lord of the troops of inferior deities among the retinue of Siva, is prayed together with the atonement and offerings of floral affords in attaining enjoyment of earthly bliss [9] (p.100) and solace from all deterrents. There are a variety of legends reckoned for his acquisition of elephant heads. Popular among all is that an abiogenetic from the clay of grime by Parvathi as her descendant and was at the wrath of Siva when a challenge to enter his house. Upon beheading Ganesa, his social father Siva condoles Parvati and transplants an elephant capitulum [10] (p.111). Artificial life as an investigation was said to have been performed twice on a single azoic, which is inquisitive to the scientific authority of parthenogenesis. Adi Shankara[10] depicted Ganesa is an unborn and formless *(ajam nirvikalpam nirākāram ekam)*, which may be the superlative of faith that is always fastened to spiritual divinity called supernatural power *(māya)*. The nature of *māya* may be hypothesized with modern scientific appreciation as an appropriation of judgement through the human cerebral to the comprehension in AI [11] (pp.562-568). *Taittiriya Upaniṣad[11]* explores the innermost cogitation of human consciousness, which is conceived of multiplicity in assimilation through sensory, emotions, genius, and attitudes. First *Prapāṭhaka[12]* in *Chandogya Upaniṣad* explicit that the bodily entity *(udgitha)* as the soul and God revered *udgitha* as the form for attaining life *(prāṇa)*, the vitality of breath and life-principle, and these are deities *(devtva)* inside man and body organs of senses. By virtue, it is the conglomerate of monumentality in organics across immense formulate and prospects of cloning on a volatile and alien system must be investigated to be a reality or deceitful.

Ancient myths of Greeks were no different than Hindus and with timeless aspirations about artificial life and immortality. Folklores of mythological heroes of Hephaestus and Daedalus integrated the idea of artificial beings in the form of intelligent robots like Pandora. Hephaestus[13] contrived all the weapons of the gods in Olympus, and Daedalus[14] created Icarus with artificial wings and who flew very close to the solar flare, and the wax-built wings melted due to heat, and Icarus fell to Earth to his death. Talos and Pandora were also the mystique of invention by Hephaestus. Ancient Greeks had been very thoughtful on the perception of biotechne[15], on how organics of human outgrowth such as aging can be challenged with technology. Scholars contest Greek antiquity about the truth or mythical about automata. Heron of Alexandria[16] did fabricate many originations of speech automata. Superstitiously, "Pandora was a kind of AI agent, who was bestowed with gods' knowledge by Hephaestus, and the quest was to pervade the human agility by exoneration of her jar of miseries." [12]. Greek visual-artists and authors described amazement *(sebas)*, wonder *(thauma)*, and astonishment *(thambos)* about these obscure memoirs. Homer[17] verbalizes those automata "of their own motion they entered the conclave of Gods on Olympus", may have ascended out of complacence that 'administrators do not require any aide' and 'masters about their slaves'. Aristotle[18] speculated in his Politics[19] that automata could, in the future, for instance, bring about human tolerance and eradication of slavery. This prophetic vision sounds a valid pretext in thoughtful to build AgI, and to be reduced to appropriate automata in elevating human ambitions alone but not the utopia of AsI.

Egyptians burdened to have forged automatons for not only by multifarious shapes but also motion and voice to the idols. An Egyptian legend of Rocail[20], while relinquishing the giant's realm and after attaining dignity and honor, fabricated a magnificent palace and a sepulcher embodying autonomous statuette discovered by the legion for blessings, and they were mistaken for having souls [13]. Gaston Maspero[21] was vocal about Egyptians speaking marvels of their deities, fabricated with painted cosmetics and chiseled wood of carved limbs, and choir mediated by temple priests, and occasionally reciprocated to questions and even conveyed eloquently conspicuous speeches. One idol in the temple of Amun in Thebes was said to move its' robotic arm and intelligently point to the next pharaoh. The priests' role modeled as an emissary between gods and mortals and had been firm believers of souls of divinities inhabited the idols and modulated in producing voices and movements to prosthetic gears and levers.

Away from robotic idols by Egyptians, Hindu philosophy identify deity *(devatva)* within a being and Ayurveda[22] maintained that the third eye is represented by the *ajna chakra*, and comparable to ancient Egypt, the symbol of the Eye of Horus *(wedjat eye)* is placing of pineal gland in portrait of human head. For humans, this tiny pineal gland at the base of the skull, and production of melatonin is highly far-reaching in research on AgI, which affects circadian rhythms and reproductive hormones. Dimethyltryptamine[23], 'the spirit molecule', may have been ignited in smaller quantities during dreams or depersonalization [14]. This mystic may be of control mechanics in humans for ages to come by AI and may not be for an inanimate to bring to life, but only to trounce human organics by dubious tactic. Mysticism[24] has dominated the ancient sages and continues to dominate the civilization even today, AgI nolens-volens may be delighted as either an invalidated and invisible god matter or golem of the forthcoming.

Analogous to the verses in *Kena Upanishad*[25] with a different argument from the above as reasoning of collective presence with fourteen reciprocal senses, inclusive of consciousness *(Ātman)*, sense of senses and the supreme sense — all these senses compete to govern organization, functioning, instrumentation of the consciousness and organic devices which are composed during its' augmentation of living matter. *"Unless animated by the intelligence of Ātman, the mind cannot perform its' functions of volition and determination. The word mind here includes both the doubting faculty (manas) and the determinative faculty (buddhi). The knowledge that of Ātman is the Eye of eye, and it is again Ātman which endows the life (prāna) with power to discharge its' functions"*. Subconscious though, the tremendous and greater compelling mind loses nothing of the senses on receipt, whether active or passive, and endures this wealth in an inexhaustible store of memory *(akṣitaṁ śravaḥ)*. Exterior of sustenance on perception may not be a direct consumer, but the subconscious mind attends, receives, treasures these sensorial inferences at an infallible accuracy [15]. Is AgI be nothing but models of engagement between manas and buddhi which is enabled through *akṣitaṁ śravaḥ*, the persisted memory and permissibility of access, and adorable of all states of mind *(tadavana)*? Where should be the admittance to *prāna* and *Ātman* in attainment of extreme artificiality? The pie in the sky is going to haunt humans for decades of panic about becoming inferior by unknown and artificial.

PERPLEXITY OF TECHNOPHOBIA[26]

"AI will either be the best thing that's ever happened to us, or it will be the worst thing. The development of full AI could spell the end of humanity. The primitive forms of AI developed so far have already proved very useful but fear the consequences of creating something that can match or surpass mankind. Humans, who are limited by slow biological evolution, couldn't compete and would be superseded. Real risk with AI isn't malice, but competence. Short-term impact of AI depends on who controls it, the long-term impact depends on whether it can be controlled at all". Stephen Hawking[27] argued terrifying proposition of doomsday and human race may have no other option to either recede to extinction or obey flunky.

MIT[28] radicalized tools that can detect incidental thoughts of hombre by wielding wearables, signals from emanating from the encephalon and pharynx be passively intercepted for further analysis. Technology companies, as proxies to Chinese military, are infusing massive attempts in programs that can permit these devices to be implantable to masses, so interfaced with highly available AI for autonomous and purposive control. China attempting trials on bio-digital social programming at Africans through infusing historical sentiments. Social programming applies emotions, culture, touch, sound, sight, voice and proximity of pertinent subject fields and exposed bio-matter through modes of performing arts to discretely reorient a person or an entire society. It had been a greatest perplex to life scientists for decades about the pineal gland in the human body, which may permit large-scale social hypnosis. AI supported administration of psychedelics[29] can produce weird reactions in cognitive processes, mood and perception — affect all human senses while altering his/ her thinking, time-sense and behavioral emotions. The secretions, by social sentiments, of biochemicals in pineal gland using AI trigger is argued to demonstrate dramatic outcome in human behaviors. In multiple steps of research, very ambitious outgrowth is expected to the ultimate progeniture of bio-engineering practice through human to animal hybrid integration plausibility. Though it is questioned of ethics by most nations, China alleged to be conducting higher level of research on injection of human stem-cells to animals prior to birth, and vice versa. Alternatively, China is also seeking in induction of genetic modification and drugs that can enhance human innate abilities of advanced senses, and intelligence and are led by deep learnt vitality through AI, and automation of interface with data protocols [16]. This publication by 'The AI Organization[30] spills the beans with a very slippery argument and draconian that does not substantiate on correlation with any scholarly research of bioengineering[31]. The discussed phraseology is structurally alien to each other and speculative with ambiguous among primacy of facial recognition, biometrics, aerial surveillance, robotics, 5G, bioengineering, AgI, digital quantum AI brain, social engineering and AsI — knowing that biological penchant of a being is very unique and not analogous to any other fellow human due to intellect and genetic variances. And descrambling and task segregation of mind signals may be a near impossibility where qualification for regularity is godforsaken.

When the ethics of an AsI is abysmally depicted, Midas[32] paradigm, they are so smartly thinking machines may surprise with how they achieve their objectives, a proportion of GIGO[33] or counterproductive. Even after the goals are predetermined, there might be dreadful aftereffects of discomfort to mankind, and these risks are explored as deriving experiments recommended by Nick Bostrom[34]. Turning to AsI, the machine might likely to stimulate egotistical knowledge of intent and desire towards sustained instinct and keep assimilating more resources with which it can achieve undesirable. Superiority with such comprehension, AsI could reconstitute itself and advance to newer and self-defined tasks, and it is virtually impossible to know the performance may not be aligned with human requisites. Additionally, the risk scenario is that any such AsI may simply be indifferent to human existence. All these risks are predicated on its masters, while exaggerated autonomy is given to machines to act in the material world. Indeed, a more pressing risk is that the current AnI is sufficiently autonomous with unsupervised learning. AsI would emerge slowly as humans painfully build better and better systems [17]. Scary though — the machine to grow progressively like human organs of tissues and the gray matter in attaining the AsI. It is obvious that the knowledge being amassed by biological and organic cognition is substantially pervasive, whereas the premise of the machines mustering inferences is through comprehension among digital errands. Theatricals of data provisioning as supremacy would persist with humans forever and be ineradicable through the progressive evolution of species. Other theatricals of movie making are not so sober, but instrumental in stimulating higher technophobia among the masses about the terror of nonliving to take over.

INCREDULOUS OF CELLULOID APOCRYPHAL

Metaphoric of conception or inception, the world of motion picture has a quantum of partage in promotion of incredible among time provoked artistry, and intense dramatics through pseudo parade of visually rich magnitude among sci-fi frolics. Be it theoretical physics[35] or illusionary myths, have been polychromatic on a magnificent canvas, encouraged by humongous global patronage.

"The search for our beginning could lead to our end", an authoritarian tagline and watchword of *Prometheus (2012)*, directed by Ridley Scott, a runaway success at the box-office. Plot revolves around elemental assumption in exposition of a clue that humans are engineered by super-intelligent morphons elsewhere in the darkest parts of the universe, which leads a team of explorers to time-travel along with two brilliant young scientists in the expedition. The inquisitive research by the revelation of an event occurred millions of years ago. A state-of-the-art humanoid alien race prefers primordial earth for their strategic activity, and disintegrates to perform the DNA seeding the planet. The spaceship's crew travels in hibernation stasis at the speed of light, while a super-intelligent android, David, pilot cruises through the entire voyage. David theoretically is incompetent of humanist emotions — such as amusement, frustration, or even desire. But the android was programmed with incident wise tendency for emulating human emotional and intellectual processes, and not of human-like consciousness [18] (pp.240-244). Beyond the descriptive rendezvous in optimism to reach a race of benevolent, godlike beings, while also debunking any spiritual notions is an adjunct logic throughout the script of roleplay as iconoclasm, the dualistic bewilderment.

"Man has made his match. Now it's his problem". *Blade Runner (1982)*, directed by Ridley Scott, imagines a world in which a group of enslaved replicants[36] revolted on another planet. Designed to live only a few years, they came to Earth to find a way to extend their life span. Building automata companions is a significant thought process for longevity but remains the exhibition of unknown and speculative exuberance of an alien AI.

"She'll keep you safe", *I Am Mother (2019)*, directed by Grant Sputore, an extinction level extermination rocks planet Earth, and an underground bunker to save embryos for future rehabilitation. A robot, Mother, commences the program and cultivates a human embryo into a baby, and raises through childhood. But their solitary bond is jeopardized when an outlander arrives from the ravaged and forbidden area in affinity to natural genetics. The plot elucidates that those emotions between natural species are boundless when compared to artificial automata that remains shallow.

"Humans and Transformers are at war. The key to saving our future lies buried in the secrets of the past". Alien cyborg robots from the planet Cybertron[37], *Transformers (2007)*, with the tagline of *"Their war. Our world."*, the first in the series from Michael Bay, created a dent with the unique identity of characters that have a captivating screenplay. They also have the unique intelligence to transform their form for skirmish, morphing into machines and mounted weapons. The destiny of humanity is at stake when two of the ideologies from robots, the godsent Autobots[38] and the nefarious Decepticons[39], bring their battle to Earth to continue their age-old feud. The robots mutate into a variety of mechanical paraphernalia as they strive to attain their eventual behest of power. Sam Witwicky, a human youth, only can salvage the world from possible destruction. *"Revenge of the Fallen (2009), Dark of the Moon (2011), Age of Extinction (2014), The Last Knight (2017)"* are the sequels on *Transformers* [19]. The history reappears about a deadly threat of resurrection from earths' history in a fight for acquisition of a lost artifact between Autobots and Decepticons. Optimus[40] discovers his dead native planet, Cybertron, while accepting that he was responsible for its annihilation. He discovers a possibility to bring Cybertron back to revival, but in support of doing so, Optimus essentially needs to find that artifact which is hidden somewhere on Earth [20] (pp.395-396). What manifests humans to fantasize of Transformers can possess minds besides hominoids? This delinquency is known to philosophers as 'the problem of other minds' — is that it seems that any action, facial expression, or language use as an exhibition by mighty machine could be nothing but the programmed response of an unthinking thing [21]. The annoying visual extravaganza, accelerated by advancement in computational graphics, showcases pseudo-AI wizardry of weird objects in composition of a legoism[41] kind.

Cinematic megalomaniacal, by using newer technologies, facilitate virtual voyage of grandiose by creative generous of visual graphic arts. This creativity causes mesmerizing imaginations of surreal and dreamlike transformative representations. Despite the superfluity of contrasting sci-fi futurist depictions about our forthcoming, unscientific to laws of physics, it psychologically feels almost difficult than ever to predict what our world will look like in the future. But, the practical applications of AI remained paused at DL, and the information industry remained mute on future developments. But research efforts are continuous, though it is very difficult to decide prior investments and future hypothesis.

CHAOTIC PROCLAMATIONS OF DIGITAL PROWESS

The first successful trial of rDNA[42] was accomplished by Paul Berg and other fellow researchers at Stanford University during the early 1970s. And this activity was instantaneously regarded as 'playing God' and the anticipation of potential biohazards posed by recombinant pathogens. In an extraordinary alarm for self-restraint, scientists in the biological discipline called for a moratorium on experimentation with rDNA [22]. The other ugly ramifications included the deployment and use of biological weapons and genetically engineered organisms with a developed resistance to antibiotics which would escape human control. The tenets were created at the Asilomar Conference on Beneficial AI[43] in 2017, and defined Asilomar AI Principles[44].

DARPA[45] remains an impressive opportunity farm for AI from an immature notion of no technological substratum into potentially affecting human lives with the most expected real-time decision-making. DARPA is focusing on AI, cognitive computing, and approaches from microelectronics to advance levels of quantum computing and neurosynaptic superlative processors in relation to how the brain performs information. The most debated controversy that DARPA has been moving away from IT related research, which it played a great role in development, even though this technology proposition is still struggling in its' bloom — we are not yet even close to AI. Licklider's[46] outstanding concept of "man-computer symbiosis"[47] was a foundational vision that requires to develop new types of computational methodologies before trying to arrive at augmented human like faculties, and a pathfinder to probable AI. The grail quest of computing may possibly be a true AI, as holy as its' perception to be the matter of electromigration[48] and ion exchange. There is no technological ambition like past by DARPA, because industry focus is shifting towards the ultimate pursuit of delivering a god-like power. Although a search for AI may ultimately be hallow and may partially be achievable as DL, even for a settlement of a minimal of symbiosis, it is a long way to go before this more limited vision is being available for human use [23]. The mutual relation may be a permanent feature as complementary to each other than the utopian extravaganza of complete surrender to robots with AgI.

"Can AI invent?" No, in accordance with numerous patent organizations and patent laws operating around the world. The AI here is a lexical comprehension of deep learned connections between phrases and sentences. The generated text for a described purpose may entirely be different from the written text, which patent laws refer to human inventors as individuals or persons, and the legal boundary of associativity to an automated text generation becomes contentious to legal interpretations. The USPTO[49] recently analyzed this auto-generated textual language by an AI cannot definitely be an inventor. Referring to a Decision on Petition, Application No. 16/524,350 (April 27, 2020), this application had been listed as no human inventor or human co-inventor, instead applied by an AI inventor identified as DABUS[50]. Since this patent application failed to declare the name as a human inventor, the U.S. Patent laws depict the policies and rules

express language for the requirement of human inventors, USPTO finally denied this patent application citing that *"only natural persons can be inventors"* [24]. Along, flippancy across the content of subject matter also must be under legal scrutiny, since the imaginative would entirely be illusionary to practicality of fabrication or programmability — artificial language on vocabulary in domain specific diction of hodgepodge, and susceptibility to assessment of legality with claims. Human description of inventions is driven by the earthly experiences of domain specific consciousness and not by all imaginative combinations of dictionary drivers.

GREGARIOUS AND PERVASIVE TRANSGENIC CONSCIOUSNESS

A humanoid or a biologically artificial with complete AgI may be furthermost intent of neo-Frankenstein, the systems experimented or developed, so far, are ridden by its construct of event-driven metaphor and have been the biggest barrier for further thought processing. Progressions in digital consciousness are in a quandary due to imperceptibility in deciphering the dimensions of human cognition to the prevailing software development practices. The differences in divergence out of the human brain with dynamically synchronized coordination to mind are the result of interpretation and responses by a centroidal executive possessorship of the prefrontal cortex. And this fusion is very organic than the known dilemma of local selectivity by any DL systems. Information meant for processing can be highly ambiguous in several ways in application to habitude, which can also greatly reduce the scope from a broader context of applicability. The coding and coordinated interactions are shared responsibilities at each molecule in physiology, and this dexterity has taken millions of years to evolve. Contextual disambiguation is ubiquitous throughout perception, and constraints are from multiple of parallel heterogeneous contexts in humans. By just being with sensory knowledge extraction, it is highly impossible to infer or interpret, since the nature of consciousnesses is multifaceted from knowledge dynamics of either prior acquired or the new learnings. Phylogenetic and ontogenetic are creative coordinated differences and may be due to the evolution and development of that specific strengths by ecology and trying to challenge any such capability is a question of whether or not. The training of AnI is restricted to only inferential of secondary information, hence the scope for noise in interpretations and confidence levels to the lowest of applicability [25].

The human psychic system is hypothesized as interwoven amalgamation across two subsets, first constituent is the dynamic blend which defines characteristics of all elements of control by corresponding thoughts. And the second, representation of productive use of thought-generating constructs as endeavors. These are made up of numerous aggregates among neurons bridged by multiple axons and dendrites, which is active assembly of emerging conformations with which the communication system can sense these by itself. The system of thought-generation will never perform on state deterministic methodology. It would be continually shaped from a coordinated assembly of active dynamic components with varying limits of potentiality and while increasingly based on acquired experiences as emerged representations. Declarative constraints in relationally multiple are of demand to arrange the constituents among associative components because it is the knowledge and experience that are responsible for the generation of thoughts. Artificial psyche in defining experiences and corresponding knowledge, such as the diction in a language dialect and the formation of comprehension, and all emotions to the linguistic composition of sentences that are understood and formed into phrases through state-based inferences. Though highly impractical, the creation of artificial mental representations and to designate as continuously learning and experiencing the formation of emotions, and which is aimed at improving its aptitude in generating the appropriate knowledge representations. The inferred experiences shall be among distributed infrastructure constituted over multiple of corporeal systems with locally synchronized artificial consciousnesses [26].

"Ten years ago, a neuroscientist said that within a decade, he could simulate a human brain. Spoiler: It didn't happen" [27]. George H. W. Bush declared that of 1990~2000 to have been the *"decade of the brain"*. It appears that decade remained with just that hype, the question of whether circumstantial of any fundamental and substantive progress. Human Brain Project started in 2009 with extravagant ballyhoo, and collaboration for swarm science across one hundred and fifty institutions around the world. Confronted with issues concerning verifiability, few years later pint-sized, shamelessly, as a software project in furnishing application tools and methods for intended scientists on important research. Human Brain Project may be contemplated as the most atrocious precedent of occult about intelligence, expected to resurrect again through 'Big Science', a highly ambitious project from Obama Administration, whose centerpieces are computing infrastructure. 'Brain Research through Advancing Innovative Neurotechnology' (BRAIN), must emphasize constituting deep knowledge for brain supportive sustenance than the hoopla of simulating an artificial brain. While the BRAIN fundings are clearly Big Science projects, and are of engineering than excitement to metaphysics, large scale neuroscience efforts, with a higher belligerence, are almost universally settled as big data works and are meant for survival through data deluge [28] (pp.246-249) among data-lake kind of applications. Rhetoric about swarm science and hive minds, finally settling to a computer-centric viewpoint of the world, where human coherence is downplayed to approval of machines. Moreover, the fanciful scientific forthcoming of artificial neuroplasticity[51] may forever be a pipedream. The dominant reason for distressing AI in incubation and in an infinite state of gestation is due to the delay in the birth of AgI, the bandwagon of brain decoding, which is not even made a serious beginning yet.

Alexandre Koyré[52] has pointed out that the greatest advancement in the scientific revolution from seventeenth century was the substitution of Aristotelian scientific philosophy by Galileo's discoveries of an abstract and ideal world. Human language faculty only enables to grasp patterns of dependent linking and the constitutive feature of systematic number assignments [29], numbers as language of measurements to real-world objects and mathematics is the grammar. Our solar system is the closest representation of such classification through natural models and measurability, Isaac Newton rephrased as *"the system of the world"*. Natural systems are somehow synchronous to each other based on deeds across relational goals, yet independent as executive, and these interactions advocated by compelling constraints and are always contextual to mission [30] *{i}*. The composed tasks are synchronous to each other in a systemic understanding of parallelism to a mission accomplishment and relationship is logical (physical models of mathematic sense) solicitation, though tasks are very independent to their nature and domain. The state management of individually parallel systems is very distinctive with no explicit relationship or inducive dependency on other parallels, and overall objective of mission is the collective obligation, like systems of nature. Contrary, dimensionality through coordination of events on the best possible DL system is suffering due to underlying intimidation by the structure of aberrant 'task hierarchies', 'process loops', 'wrapped events' *(APIs on APIs on APIs...)* and 'noisy datasets' — afflicted by the long-drawn heritage of *if... then... else... endif...* syndrome, and declarative goblins of globally reusable monoliths. All those AI progressives who have not been able to assimilate the methodology of bionomics and surrounded sensibilities, may have to give away in search of a genuine in the future.

APPERCEPTIVE TERRESTRIAL SYSTEMATIZATION

"All things come to an end", but astonishingly a meek end, proclaimed to be cautious apprehension of potentiality towards racial profiling and may causing human rights abuse [31], "IBM Watson Visual Recognition is discontinued. Existing instances are supported until 1 December 2021, but as of 7 January 2021, you can't create instances. Any instance that exists on 1 December 2021 will be deleted" [32]. All monumental passions of 'IBM Computer Vision', have come to a grinding halt in 2019. Brilliance in image recognition and cognitive discovery to serviced bitmaps of 'portraits akin recognizable person' and 'bit profile knowledge of beast to determine the type of species' is synonymous to leisure instruments of industrial revolution of Europe, best of purpose than to serve the

fancy of royals. The proficiency in applicability to subastral engineering on dimensional real-world jigsaw features and manifestation to visual cognition is a bizarre dilemma in AI. This is due to dearth in AI conception to the extremes of superlative visual cognition by humans, which is dynamically scalable by aptitude to the domain of prior subject knowledge, and gigantic proposition to any scheme of artificiality. Indeed, visual appreciation by hominid is far superior and complex than lexicon. Application published at USPTO for IBM on the similar enhancements to learning systems is done by author *{i}* [33]. An expert engineering professional observes key graphics of the source information from multiple varieties of physical sizes and relative features across, the perceptual intuitiveness triggers the faculties of higher understanding for methods solution. The sources are merged in a purpose driven supervised intrinsic understanding of virtuality in multiple observations, irrespective of variety. This phenomenon does not require any edge matching, feature extrapolation *etc.*, in the way downstream application require as source for target design activities. Human perception alike knowledge about the features and dimensions is inferred virtually in support of physical calibration to target requirements. Learning shall have multiple of supervised iterations, so that maturity of information attained against the model of application to purpose. When the model has been sufficiently refined, the system makes the knowledge available to downstream and domain engineering systems. Similar observation in learning a determined model from multiple of visual frames be possible, provided the AI is built with sufficient learning capabilities in segregation of feature profiles based on application model and build virtual canvas of visual knowledge with human like consciousness across. This variety of proficiency may remain a blocked methodology by this patent application, and most probably hang in the air forever.

CONCLUSION

Early anthropologists like Tylor[53] and Morgan[54], theorized in evolutionary terms of differentiation about interpretivism and naturalism, all historical changes and social innovations cause development of the intrinsic mind. Modularity is *"the view that many fundamentally different kinds of psychological mechanisms must be postulated in order to explain the facts of mental life"*. Predetermined symbolic capability is the primary competence in human ability to learn languages despite the deficiency of stimuli in a person's living conditions, since modular mind is surrounded by its' boundaries of motor and perceptual systems. Cognitive system is then perceived to be as modular to the extreme in a strong sense of lesser domain centric attitudes which have fixed and redundant encapsulations. Contrary, the central cortex is less modular and again not composed of domain determined sub-systems and associative encapsulations [34], but only the master who maintains their constrained logical relations. Human intelligence is multifaceted and deceptive to any form of passive acquisition and artificially active replication by an external arbiter, because the natural receptive systems are based on the motivation of task completion. In case the task is strenuous or perceptibly insurmountable, the mind plays a quick ABTS[55], and returns all the cognitive faculties to normal state. The artificial may not ascertain such a phenomenon and remains the victim of Gödel's[56] quintessential loop of system exceptions of progressive irrationality. This incompleteness may well be due to the foundational lapse of methodically inferential with arithmetization of communicative languages. The unknown of AgI of being with capability as an obedient interpreter, shall mature to the levels of natural understanding in communications while using diagonalization with contextual trigger of speeds in comprehension by persisted cortical knowledge. Backprop[57] has been the triumph by DL so far, if progress is expected on the way forward for an AI, then dependency on Backprop must shift away to a much wiser methodology. Assimilation of knowledge in human mind is much different than that of intelligent machines are being thought of today. Much serious investigation for alternative techniques is inevitable [35]. It is too early to judge the egomania of new methods of AutoDL[58], AutoML[59] and Neural Architecture Search (NAS)[60] — a semblance to avoid knowledge and interactions atypical to humans, and towards a fancy legoism to a whimsical world of Autobots and Decepticons. *IBM Deep Blue* made history to beat a world champion *Garry Kasparov* (1997)! *IBM Watson* surpassed world's best players on the popular quiz show *Jeopardy* (2011)! Did these events hoodwink entire humanity to a flamboyant disposition? Significantly, both IBM Watson Health and Alphabet DeepMind flunk-off market confidence. IBM has burdened by *"a fundamental mismatch between the way machines learn and the way doctors work"* [36]. DeepMind has ascertained that *"what works for Go may not work for the challenging problems that DeepMind aspires to solve with AI"* [37]. Will there be a possibility of total dereliction to established methods and renaissance from the ashes to amorphic rendition in AgI? This brings us back to the ideologically ambitious theories of AI, AgI and AsI to interrogations of ingenuity in contesting against human consciousness, which is the dominion of bionomics with deep intuitions across sensory and mind coordination, the truth to Newtonian vision of *"the system of the world"*.

Candid Exhortation

Instead of AI chimeric, substantiate the human mind with more prowess, which is struggling the classic design deficiency in hominid system, of centralized control with coexistence of equidistributional capillaries. Overwhelmed DL systems may be, as mind support system to technophilia[61], of simulating potentiality to the context specific needs of organic exigencies for brain and body of humans, and temporal inferences by pathological lesson that is explicit to very discrete as subject isolation. This is better objective, where the inferential assessment to plasmatic redressal may be consummate to longevity or (probably) singularity[62]. Addressing the enzymatic distress and appropriate organic acclimatization is a highly personalized therapeutic inducement of pharmaceutical substances, possibly either oral admission or intravenous, depending on the importunity of situation [38] (p.196). It may not be that obvious, but DL with inferential knowledge of mineral composites and ability to simulate the process models of drug composites for timely body alchemy looks, possibly, a categorical accomplishment by artificial learning — but subject to higher research in organic chemistry, biotechnology, biochemistry, social ethics, and legal authorizations.

NOTES

[1] The Liar paradox was discussed for more than 2,000 years ago, and multiple inferences of true and false coexist among every kind of physical cognition.

[2] John Locke (1632~1704) was an English philosopher and physician, and widely observed as one of the most influential thinkers of enlightenment. He holds that human is a matter of psychological continuity and considered it to be founded through consciousness and experience.

[3] Mary Wollstonecraft Shelley (1797~1851) write those famous novels of *Frankenstein or The Modern Prometheus* and *The Last Man* and pressed a modern political constrict into science fiction concerned on ethical embarrassment of an artificial life, and probable through science, technology, and other forms of cultural dynamism.

[4] William Godwin (1756~1836) was a dominant contributor to radicalism during the period of Romantic movement, and provided with the central theme that man, once released from the burden of all artificially political and social constraints, would stand in absolute rational harmony with the world.

[5] Christopher Langton is a computing scientist named the field of Artificial Life during 1987 and organized a first International Conference on "The Synthesis and Simulation of Living Systems", and a concept that brought together the concerned who are interested in building the computing models.

[6] Alan Mathison Turing (1912~1954) is universally considered as father of the theoretical computer science and artificial intelligence. formalised the concepts of

algorithm and computation with the Turing machine and considered as model of a general-purpose computing machine.

[7] Prometheus is a Titan god of fire, and best known for defying other gods while stealing fire from them and generously distributing it to humanity in enrich them with technology, knowledge, and more profoundly, the civilization.

[8] Dialectical Materialism is a philosophy of rationalism to nature, history, and science which was researched by Karl Marx and Friedrich Engels.

[9] Georg W. Friedrich Hegel (1770~1831) was a German philosopher. He was considered as one of the most important personalities in German idealism and one of the founding researchers in modern and Western philosophy.

[10] Adi Shankara (8th cent. CE) was a Hindu *Vedic* scholar and teacher. His works present a harmonizing recital of the sacred scriptures. The theme is to liberate knowledge of self at its' core, while synthesizing to *Advaita Vedanta* teachings of his time.

[11] *Upanishads* are sacred treatises explained in *Sanskrit* c.800~200 BC, rationalizing the *Vedas* in essentially in mystical and monistic sense.

[12] *Prapāṭhaka* is a main division of a book or treatise, a chapter.

[13] Hephaestus was the Greek god of blacksmiths, metalworking, crafts, technology, the forge and fire.

[14] Daedalus was a very proficient architect and a fabricator too. He was also identified as a symbol of power, knowledge and wisdom.

[15] Classical stories describing Greeks might have termed biotechne (bios, life, and techne), and preternatural in optimism about biotechnology.

[16] Hero of Alexandria (010~070) was a Greek engineer and mathematician who was from city of Alexandria in Roman Egypt. He was often considered as a greatest pathfinder of antiquity from the Hellenistic scientific tradition.

[17] Homer, a fictive poet, if he ever was a domineer said to have enamoured 'Iliad' and 'Odyssey'.

[18] Aristotle (384~322BC) was a Greek polymath and philosopher during the high of Classical period in Ancient Greece, and a disciple of Plato.

[19] Aristotle observes human beings are creatures of flesh and blood — 'Man is a political animal', rubbing shoulders with each other in communities. "Politics" was a publication of political philosophy by Aristotle, and divided into eight books.

[20] Rocail was the younger brother of Egyptian deity Seth, the third son of the biblical Adam and Eve.

[21] Sir Gaston Camille Charles Maspero (1846~1916) was a French Egyptologist famous for his contributions of Egyptian hieroglyphics, and instrumental in developing Institut français d'archéologie orientale and director-general of excavations and of the antiquities of Egypt.

[22] Ayurveda is believed to be a bestowed medical knowledge from the gods to sages, and then to human physicians.

[23] N,N-Dimethyltryptamine (DMT) is a substituted tryptamine which is both a derivative and a structural analogue of tryptamine.

[24] According to Thayer's Greek Lexicon, the term mysticism in classical Greek meant "a hidden thing" or "secrecy".

[25] Kena Upanishad questions the ecology of man, and his/ her origins, essence into relationship of him/ her with knowledge and larger design of sensory perception.

[26] Technophobia is also popularly known as technofear, and is the dislike or fear of advancement in technology or admixture of devices, especially computers. While technology continues to evolve, the diversity of technophobia becomes more complex. Sometimes technophobia is also interpreted in the impression of an irrational fear, but there is a justified consensus on such fears.

[27] Stephen William Hawking (1942~2018) was a cosmologist, theoretical physicist, and authored the book '*A Brief History of Time*'.

[28] Massachusetts Institute of Technology (MIT) is a land-grant research university in Cambridge, Massachusetts. MIT has been playing a pioneering role in the research of modern technology and science, which ranks it as one of the top academic institutions of the world.

[29] Class of psychoactive substances — some transpire naturally through fungi, seeds, and leaves; breweries of vines; and others are artificially produced in laboratories.

[30] Mission of the AI Organization, the to "expose emerging threats to people via AI, robotics, and bio-digital social programming, and lead in safeguarding human values rather than AI, and strongly rooted by principle and believes in keeping humans first and machines last".

[31] Bioengineering applies the principles of biology to the preparation of engineering tools for viable, tangible and usable products. The subject specialization engages knowledge and competence from pure and applied sciences, like mass and heat transfer, kinetics, biomechanics, biocatalysts, bioinformatics, purification and separation processes, surface and polymer science, fluid mechanics, thermodynamics, bioreactor design.

[32] Midas is a Greek and Roman legend, also known as king of Phrygia, known for his greed and foolishness.

[33] The phrase of "Garbage In and Garbage Out" was first used during 1951. US Army mathematicians from their work with the very early computers where William D. Mellin derived a phenomenon of that the computers cannot judge for themselves, and the programmes that are "sloppy programmed" would lead to inconsistent outputs.

[34] Nick Bostrom, University of Oxford, is a philosopher known for the work of anthropic principle, existential risk, human enactment ethics, reversal test and AsI risks. And to study the impacts of future technologies, he founded "Oxford Martin Program".

[35] Theoretical physics, a branch of physics, demonstrates mathematical models surrounding the abstractions to rationalization of physical systems while predicting the phenomenon occurring naturally. Contrary, empirical physics applies experimental tools to make the natural phenomenon is proved for the truth to originality of occurrence.

[36] Replicants are genetically engineered human-like creatures.

[37] Cybertron is the fictious home planet of the Transformers and the body of their creator is Primus.

[38] Autobots are highly benevolent, apperceptive and self-configuring robotic-forms from the planet of Cybertron.

[39] Decepticons are the main antagonists, often superior and powerful vehicles, that are in the shapes of military vehicles, construction vehicles, aircrafts, sports cars and even mutated smaller sized objects.

[40] Optimus, also known as Orion Pax, is the main protagonist and the last Prime from Transformers film series story and chieftain of the Autobots, and also the brother of Megatron, who is the leader of Decepticons.

[41] Legoism *{1}*, free connotation to toy-culture in application of weird multi-dimensional in debilitated architectural eccentric forms, surrealistic urban design, bizarre product designing, jaunty celluloid storytelling, *etc.*

[42] rDNA is a type of unique genetic engineering that engages the splicing of genes into alternative organisms.

[43] Conference was organized by "Future of Life Institute".

[44] Asilomar AI Principles are subdivided broadly into three primary categories of among Ethics and Values, Research, and Longer-Term Issues.

[45] For sixty years, "Defence Advanced Research Projects Agency" has held focused mission which made pivotal investments in development of breakthrough

technologies required for US national security.

[46] Joseph Carl Robnett Licklider (1915~1990) was an American psychologist and also a computer scientist who is recognized to be among the most prominent personalities in development of computer science and general computing.

[47] Something of Lickliders' vision for a symbiotic relationship between humans and computers at a potential time of the future.

[48] Electromigration is the movement of atoms based on the flow of current through a material.

[49] "The United States Patent and Trademark Office" is the federal bureau for officially granting U.S. patents and registry of trademarks.

[50] "Device for the Autonomous Bootingstraiming of Unified Sentience", DABUS created by Stephen Thaler as a text generation AI, and the system anticipated to simulate human brainstorming subject matter in creation of new inventions.

[51] The nervous system has the ability to change the course of activity with reference to intrinsic or extrinsic impetus by reorganizing its' functions, structure, and corresponding connections.

[52] Alexandr Vladimirovich Koyra (1892~1964) was a philosopher who wrote on the philosophic history of science and metaphysics.

[53] Sir Edward Burnett Tylor (1832~1917) was an English anthropologist, and founder of cultural anthropology while his ideas typify "nineteenth century cultural evolutionism".

[54] Lewis Henry Morgan (1818~1881) was a pioneering American social theorist and anthropologist, and widely known for his work on "kinship and social structure". His other theory on social evolution is "ethnography of the Iroquois".

[55] 'Abandon the Ship' is a unique capability of highly judged ability to the evolved species.

[56] "Gödel's incompleteness theorems" showcase important results among modern logic and derive deep computing implications on various concerning issues of model evolutions and possible failures.

[57] In machine learning and deep learning, backpropagation algorithm (aka. Supervised learning algorithm) is a popular technique in training the feedforward on neural networks.

[58] Building deep learning models may take a lot of time, but not with AutoDL, and automate meta-work so that DL knowledge on self-application of contextual models.

[59] AutoML provides both ML to use DL knowledge supported self-recovering models, make predictions, and test business scenarios.

[60] NAS strategy is to define the search space for the target neural network, deep learning model to be composed in a range of convolutional and fully connected layers. And promotes additional strategy of reinforcement learning where the DL must again iterate to optimum in the given context of problem solving.

[61] Technophilia is referred to strongest possible interest for adapting to technology, especially technologies due to automation as mobile phones, the Internet, personal computers, and home cinema — precisely opposite phenomenon to technophobia.

[62] "The Coming Technological Singularity (993)", an essay by Vernor Vinge popularized the conception of singularity as an unusual distinction. And also, being interpreted for interventional advancement for achieving human immortality.

CONSENT FOR PUBLICATON

I, K. Ravi Kumar Reddy, on behalf of all other authors of this paper, give my consent for the publication of identifiable details, which can include photograph(s) and/or videos and/or case history and/or details within the text ("Material") to be published in this chapter.

REFERENCES

[1]   S. Poole, "Art and design, the impossible world of mc escher", Available from: https://www.theguardian.com/artanddesign/2015/jun/20/the-impossible-world-of-mc-escher

[2]   D.R. Hofstadter, *I am a Strange Loop.* Perseus Books Group: Cambridge, Massachusetts, 2007, pp. 1-432.

[3]   L. Goldstein, "Reflexivity, Contradiction, Paradox and M. C. Escher", *Leonardo,* vol. 29, no. 4, pp. 299-308, 1996.
[http://dx.doi.org/10.2307/1576313]

[4]   R. David, G. Michael, and J. Beall, "Liar Paradox .Stanford Encyclopedia of Philosophy", Available from: https://plato.stanford.edu/entries/liar-paradox/

[5]   E. H. Botting, *Artificial Life After Frankenstein.* University of pennsylvania Press: Philadelphia, Pennsylvania, 2021.

[6]   J. Copeland, *The Essential Turing.* Oxford University Press Inc.: New York, 2004.
[http://dx.doi.org/10.1093/oso/9780198250791.001.0001]

[7]   A.M. Turing, "Computing machinery and intelligence", *Mind,* vol. LIX, no. 236, pp. 433-460, 1950.
[http://dx.doi.org/10.1093/mind/LIX.236.433]

[8]   "Genesis 2:7", Available from: https://www.kingjamesbibleonline.org/Genesis-Chapter-2/

[9]   J.L. Shastri, *The Shiva Purana.* vol. 1. Motilal Banarsidass Publishing House: Delhi, 1970.

[10]   J. Dawson, *A Classical Dictionary of Hindu Mythology and Religion* D K Printworld (P) Ltd: New Delhi, 2000.

[11]   V. Ramabrahmam, "A modern scientific awareness of upanishadic wisdom: implications to physiological psychology and artificial intelligence", *Proceedings of the World Congress on Vedic Sciences.* Bangalore, pp. 562-568, 2004.

[12]   A. Mayor, *Gods and Robots: Myths, Machines, and Ancient Dreams of Technology.* Princeton University Press: Princeton, New Jersey, 2018.
[http://dx.doi.org/10.2307/j.ctvc779xn]

[13]   B. Herbelot, *Bibliotheque Oriental.* Bibliothecae Electroralis Monacensis: Paris, 1697.

[14]   G. Bogdan Alexandru, R. Ioana Andreea, and C. Maria, "The morphological and functional characteristics of the pineal gland", *Med Pharm Rep.,* vol. 92, no. 3, pp. 226-234, 2019. Available from: https://www.ncbi.nlm.nih.gov/pmc/articles/PMC6709953/

[15]   Swami Nikhilananda, The Upanishads (Volume I), Katha, Isa, Kena, and Mundaka, New York, NY: Harper & Brothers Publishers, 1949.

[16]   C.A. Parsa, *Artificial Intelligence: Dangers to Humanity.* The AI Organization: California, 2019.

[17]   T. Walsh, *Machines that Think: The Future of Artificial Intelligence.* Prometheus Books: Amherst, NY, 2018.

[18]   K.D. Killian, "Prometheus", *J. Fem. Fam. Ther.,* vol. 26, no. 4, pp. 240-244, 2014.
[http://dx.doi.org/10.1080/08952833.2014.967145]

[19]   B. Stamets, "Transformers: The last knight: Securing homelands and unearthing secrets", Available from: https://www.researchgate.net/publication/318283594

[20]   C. Barsanti, *The Sci-Fi Movie Guilde, The Universe of Film from Alien to Zardoz.* Visible Ink Press: Canton, Michigan, 2015.

[21]   Ed. Shook, *Transformers and Philosophy, More than Meets the Mind.* Open Court Publishing: Illinois, Peru, 2009.

[22]   J. Kozubek, *Modern Prometheus, Editing the Human Genome with Crispr-Cas9.* CB2 8BS:

Cambridge University Press: Cambridge, 2018.

[23] *Bonvillian, at al., Eds., The DARPA Model for Transformative Technologies, Perspectives on the U.S. Defense Advanced Research Projects Agency* CB2 1ST: Open Book Publishers: Cambridge, 2019.

[24] R.N. Phelan, "Can an Artificial Intelligence (AI) be an Inventor?", Available from: https://www.patentnext.com/2021/03/can-an-artificial-intelligence-ai-be-an-inventor/

[25] A. William, "Phillips, Christoph von der Malsburg, Wolf Singer, "Dynamic Coordination in Brain and Mind", In: *Dynamic Coordination in Brain and Mind, From Neurons to Mind.* The MIT Press: Cambridge, MA, 2010, pp. 1-24.

[26] A. Cardon, *Beyond Artificial Intelligence, From Human Consciousness to Artificial Consciousness.* John Wiley & Sons: Hoboken, NJ, 2018.
[http://dx.doi.org/10.1002/9781119550983]

[27] E. Yong, "The Human Brain Project Hasn't Lived Up to Its Promise", Available from: https://www.theatlantic.com/science/archive/2019/07/ten-years-human-brain-project-simulation-mark-ram-ted-talk/594493/

[28] E.J. Larson, *The Myth of Artificial Intelligence, Why Computers Can't Think the Way We Do.* The Belknap Press of Harvard University Press: Cambridge, MA, 2021.

[29] H. Wiese, *Numbers, Language, and the Human Mind* CB2 2RU: Cambridge University Press: Cambridge, 2003.

[30] K. Ravi Kumar Reddy, "Constrained synchronous parallelism, the longevity of asymmetrically collaborative systems", *International Conference on Electrical, Computer and Communication Technologies.* Coimbatore, pp. 1-6, 2019.

[31] W. Knight, "IBM's withdrawal won't mean the end of facial recognition", Available from: https://www.wired.com/story/ibm-withdrawal-wont-mean-end-facial-recognition/

[32] I.B.M. App Connect, "IBM watson visual recognition account details", Available from: https://www.ibm.com/docs/SSTTDS_11.0.0/com.ibm.ace.icp.doc/localconn_ibmwatsonvr.html

[33] R.K. Reddy, *"Cognitive analytics for graphical legacy documents",* US 2020/0327429A1

[34] J.A. Fodor, *The Modularity of Mind.* The MIT Press: Cambridge, MA, 1983.
[http://dx.doi.org/10.7551/mitpress/4737.001.0001]

[35] S. Oman, "Deep Learning Dead-end?", Available from: https://aimatters.wordpress.com/2017/09/17/deep-learning-dead-end/

[36] E. Strickland, "How IBM watson overpromised and underdelivered on AI health care", *IEEE Spectr.,* pp. 24-31, 2019.
[http://dx.doi.org/10.1109/MSPEC.2019.8678513]

[37] G. Marcus, "Deepmind's losses and the future of artificial intelligence", Available from: https://www.wired.com/story/deepminds-losses-future-artificial-intelligence/

[38] R. Kurzweil, *The Singularity is Near, When Humans Transcend Biology* NA: Penguin Group Inc.: New York, 2005.

CHAPTER 2

Applications of Artificial Intelligence in Robotics

Pingili Sravya[1,*], **Hemachandran K.**[2] and **Ezendu Ariwa**[3]

[1] *School of Business, Woxsen University, Hyderabad, India*

[2] *Department of Artificial Intelligence, School of Business, Woxsen University, Hyderabad, India*

[3] *University of Wales Trinity Saint David, London Campus, United Kingdom*

Abstract: Artificial Intelligence is a theory of the cognitive perspective in the province with robotics to human communication with the perception of action. The ability to develop computer systems would require human intelligence to perform tasks [1]. Artificial Intelligence plays a prominent role in robotics in providing effective analytical business solutions like human behavior in the real world. The common root of artificial intelligence and robotics has a scientific interaction that transforms technological improvement in robotics application and utilization and has a potential for future robotics in various applications and AI technologies. The study of the creation of intelligent robots in Artificial Intelligence is an entity for different objectives and applications. It is known to many people that artificial intelligence is a subset of robotics. Robots have human-like behavior by which they can perform tasks like a human if enabled with Artificial Intelligence.

Keywords: Artificial Intelligence, Human Intelligence and Machine Learning, Robotics.

INTRODUCTION

Artificial Intelligence deals with machines like humans and performs human-like activities creating devices that enable machines to sense and act like humans. In the branch of technology, artificial intelligence plays a prominent role that deals with the study of robots and helps in designing intelligent machines to make life easier for everyone in everyday life. The association between the two is the coherence of computers and science, which synthesize the fact of finding brain waves to perform different operations. There is no distinction between artificial intelligence and robotics, which naturally leads the way to "intelligent machine". We might agree with the fact that not every machine is a robot; that is invented to design robots that do not deal with virtual agents.

* **Corresponding author Pingili Sravya:** School of Business, Woxsen University, Hyderabad, India;
E-mail: Sravya.pingili_2023@woxsen.edu.in

Hemachandran K., Raul V. Rodriguez, Umashankar Subramaniam & Valentina Emilia Balas (Eds.)

The vision of artificial intelligence for most of us today is that machines are not in danger of being destroyed. But AI is a real risk to the survival of human civilization. Whether artificial intelligence poses a threat is a matter of debate. Artificial Intelligence is the science and engineering of creating intelligent machines. In some ways, AI is the technology that makes devices function and behave like humans. More recently, AI has made this possible by developing apparatus and robots in many areas, such as healthcare robotics, marketing business analytics, and more. However, many applications are not recognized as AI. This is because when we think of artificial Intelligence, we often think of robots that run our daily lives. We always use truth in our daily lives because facts permeate our everyday life [2]. For example, how Google provides accurate search results and Facebook feeds consistently provide content based on our interests are artificial responses. Intelligence.

A standard misunderstanding is that numerous individuals think that artificial intelligence, machine learning, and deep learning are the same because they have typical applications. For example, Siri is his AI machine learning application, but how are these technologies erased? On the other hand, deep learning is a subset of machine learning that uses neural networks to solve complex problems. Machine learning and deep learning require artificial intelligence, which provides a set of algorithms and neural networks to solve data-driven problems. But AI isn't just banned in machine learning and deep learning. It covers many areas, such as natural language processing, object recognition, computational imagination, and robotic expert systems.

DIFFERENCE BETWEEN AI AND ROBOTICS

Futurist of robotics and artificial intelligence helps companies understand the latest technology trends in synthetic intelligence and robotics; they both are significant trends for companies to watch, but sometimes there is an overlap between them. The area of computer science helps us to materialize computer agendas that can comprehend themselves either by foraging information or operating detectors and infusions to help algorithms understand themselves. Manufacturing or building cars would take a long time and means to enroll, but today advanced robots can do almost anything. This is where the overlap between these two areas comes into play. You can combine artificial intelligence and robotics. A robot is a body, and artificial brilliance is a brainiac.

In the past, if we had robots for long-time building things like cars, they would have picked up something programmed to screw something, but they couldn't intelligently make decisions. Nowadays, we can give those robots things like cameras that act as their eyes and add our intelligence to this equation as the brain. Suddenly, artificially intelligent robots were born, and all advances in robotics were taken out of them. Think of drones as robots these days. For example, an artificial intelligence brain allows this drone to fly autonomously. Now, self-driving cars that combine robotics and brains are back on the scene. By combining artificial Intelligence and robotics, we can now have a machine vision camera to detect where the objects are and then uses its sensors to pick them up and ideally place them.

It shows how far autonomous robots have come in the meantime. It also has a feature called reinforcement learning. In other words, instead of feeding machines tons of data to perceive things. Recently we can see robots that can monitor the environment and learn to start to walk, and each step was a success for a robot. Combining AI and robotics is about data science that learns by themselves robots about machines that can do autonomous things. Still, by combining them, AI becomes the brain, the robot becomes the body, and you can achieve amazing things.

PERCEPTION AND INTELLIGENT ROBOTS

A particular task related to robot perception is the thought of the human to interact with the environment with different parts like the body, ears, nose, and eyes, *i.e.,* something about the external environment similar to that of robots for real-world understanding. Manipulating and executing operations in the real world is known as robot perception. Robot perception is the same as humans use different parts of our body; robots use various sensors [3]. From Fig. (**1**), action perception is common in robotic platforms, which is confident that interacting with the world to get sensor inputs from the real world is known as sensing and perception.

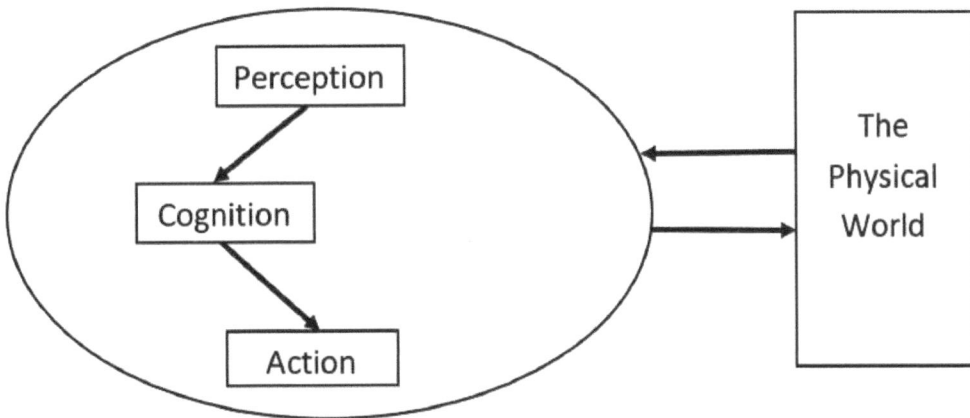

Fig. (1). Perception of robots to the physical world.

In robotics and artificial intelligence, robot perception is a prominent research field in that observable perception approaches have determined current robotic systems. Robots have to use sensors, which decide the function of actuators like arms and motors, which delivers the action so that it decides its action and sets good touch by sensing the environment, also known as action perception. Robots use different sensors, such as laser range and sonar, that find vision difficulties in undeveloped atmospheres.

A robotic representative in the naturalistic earth has to haggle with undeveloped atmospheres like vehicles. There can be static objects, so the robot should carefully decide the path without hitting anything and even consider the regulations such as traffic lights. So, perception is very complex in such an environment, so you use different sensors. For a robot to accomplish its tasks, it must endow with capabilities to interpret the behavior of dynamic and surrounding objects and agents.

AI AND ADVANCED ROBOTICS TECHNIQUES

Robots are essential in improving efficiency, productivity and, most importantly, safety despite taking people's jobs. Advanced humanoid industrial and assistance robots to change destiny with the usefulness of unnatural intelligence meet digit simulated by agility robotics integer generated to take supervision of individuals in their residences [4].

Industrial robots have achieved the next echelon of industrialization with greater flexibility in the exhibition and more immediate programming times. Artificial Intelligence (AI) and machine learning (ML) are technologies that encourage robots to accomplish numerous previously impossible tasks. Construction, validating and deploying these technologies for progressive robotic systems stay a significant challenge.

Robots and artificial intelligence are becoming so popular now that they can almost be seen in many different areas of life. There are robots for production, manufacturing, logistics, phone services, intelligent automation and many more. There are so many kinds of robots you can see everywhere worldwide, and maybe very shortly, many people in the world could be working alongside robots, or they could always be on our side to serve us whenever we need their assistance.

Within a few years, robots and intellectual mechanisms will eradicate many positions in customer assistance, truck, and taxicab kindnesses, directing to heightened engagement. According to the information, intelligent representatives conversing with bots and digital subsidiaries like Amazon's Alexa, Apple's Siri, Google Now and Facebook's Messenger BOTS are powered by artificial intelligence. You can comprehend people's demeanor, investigate their appetites, and make determinations.

Will Robots Replace the Human Workforce?

Despite taking people's appointments, robots have lived essentially in enhancing efficiency, productivity and workers' protection. The advanced humanoid industry uses artificial intelligence to change the future, accommodate a multitude of agility robots, take care of people at home, assist in disaster relief, and deliver packages designed for delivery and service robots. Retail and financial industries have recently adopted sociable humanoid robots that identify countenances and fundamental moral sentiments. All these movements display human-level skillfulness, so the robot can be flawless for examination and retrieval procedures and complete mortal tasks in circumstances where humans cannot persist. An advanced humanoid robot was developed to function autonomously and carry out severe endeavors in treacherous conditions.

Robots And Our Developing Environment

State-of-the-art humanoid industrial and assistance robots are transforming fortune with the help of artificial intelligence, corresponding to the numerals assembled by agility robotics. A robot is simply a device. Applications and software developed by people handle them. They should help us with our assignments, not take over ours entirely. Apparatuses and robots make our jobs considerably more manageable, but we exclusively utilize robotic technology when we need their services. They accomplished retaining the capability to substitute us in the workplace. Robots cannot nudge humans out of the workplace.

Robots can sense what we as humans cannot perceive, and robots belong to serving humanity for nature because deploying robots into the heart is where we can learn the most. We can witness people and robots working together in ways that can help solve environmental problems, which helps us get climate change under control through the interaction of people, robots, and our natural world. The pervasive problem we are facing of pollution everywhere can harm ecosystems.

AI, Robots *Vs* Humans

The development of robotics technology has made great strides in the last two decades. The robot is one of man's most gorgeous innovations. Scientists and individuals involved in robotics understanding are striving to assemble better interactive and supportive robots. But part of humanity worries about a dystopian fortune ruled by AI. Humans are imaginative and can come up with opinions when they produce. Unlike robots that intercept or forget if a wire comes off, humans can improve specialties and preserve functioning; robots do averse. If the wire is trimmed, the robot resolves not to reposition.

Robots have augmented intelligence, though they intend not to substitute humans. The mechanization of willpower significantly changes our lives. Job types are transformed and performed by artificial intelligence. Nevertheless, mechanization intention directs to more current jobs in various classifications, with humans over the bots to manage their work. Robots include intelligent software, but they hold customizations that most patrons ought and mandate human contact. Robots have Boolean capabilities and cannot conduct the elaborateness of mortal associations.

Humans will work with robots; they will not be renounced independently to endure specialties. Robots do not comprehend artistic standards or vernacular. They are designed with AI technology to understand a specific language spoken in a particular way. AI in robots cannot consider the context. You do not understand irrational thoughts. Not all jobs that require emotional support can be handled. For example, speaking with a customer care executive who goes to an IVR machine with limited answers, just like bots, we can say that robots are not quite efficient for customer service. Problem-solving requires an innovative technique, and robots stand not indicated to embark on such annoyances. Humans design robots established on the style of assistance or approval they ought. Robots are substitutes for distinguishing works like driving a car, grooming the house, or directing traffic.

ROBOTICS & AI FUTURE OF HUMANITY?

Some things resemble humans with surprising capabilities we've never seen before, including a form of Intelligence by doing a task the system has not performed as well as it could learn and change to do better next time. Adapting new technologies changing the world and how they will transform our lives, humans will be slowly reduced to providing human expertise. Everything else will be automated; robots can do so much for us already and more in the future [5].

In a new world where robots are among us, artificial Intelligence is our new ally, vehicles have no human driver, and the border between man and machine slowly disappears. We have entered the era of the social robot, where they left the factories and the labs. They are no longer confined to simple repetitive tasks [6]. Today these creatures of metal and plastic have permeated almost all areas of our lives. You can find them in the streets, schools in the heart of our homes and even in concert halls. Modern machines can collaborate and communicate with us, no longer considered simple tools. Despite their playful appearance, these little automatons are not just novelties. Shortly they will participate with us in daily life.

A. Robots in School

The idea seems improbable, yet the results with students are promising. Kids appreciate this little teaching assistant who engages them through play and repetition activities in the classroom that require routine interactions. Robotics specialists have persuaded these experiences will become a more significant part of our future and are essential to a 21st-century education.

B. Robots in Health Care

Tested in hospitals and retirement homes to help guide patients to identify faces and emotions and adapt their responses to the person they communicate with. In an older adult's life, it helps to memorize the usual bedtime and offers a gentle reminder to go to bed earlier [7]. From Fig. (2), we can see the life cycle of health care in robotics, that can make a difference. Knowing there is someone in the room who can be of service who can decide to bring a drink and render small kindnesses, the human-to-robot relationship could resemble a human-to-human connection.

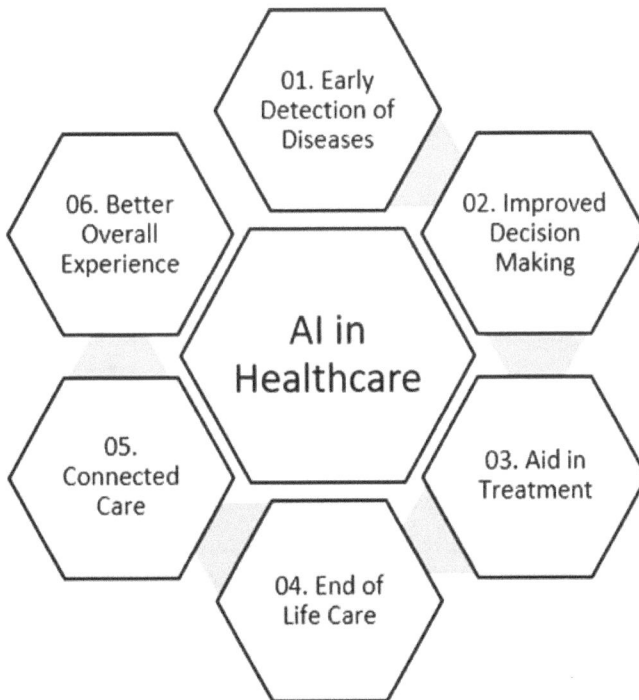

Fig. (2). AI life cycle in healthcare.

C. Robots to Analyze Emotions

Representatives of a new generation of companion robots, tiny, automated beings specifically developed to form relationships with humans. They could detect the feelings by analyzing the tone. A humanoid can be like a human to deal with inherently.

D. Robots in Industries

In real-life conditions, grabbing and moving objects is dangerous in buildings or factories for human rescuers. New generations of autonomous cars are plugged with sensors. Video cameras detect traffic lights, read signs, and even distinguish between pedestrians and bicycles. Radar and lasers pinpoint surrounding objects.

E. Robots in the Aviation Industry

Robots play an essential role in aerospace applications (Fig. **3**), like building aircraft engines and performing work such as drilling holes and painting the aircraft. Robots' reliability, performance and accuracy have made them increasingly popular in the aerospace industry [8].

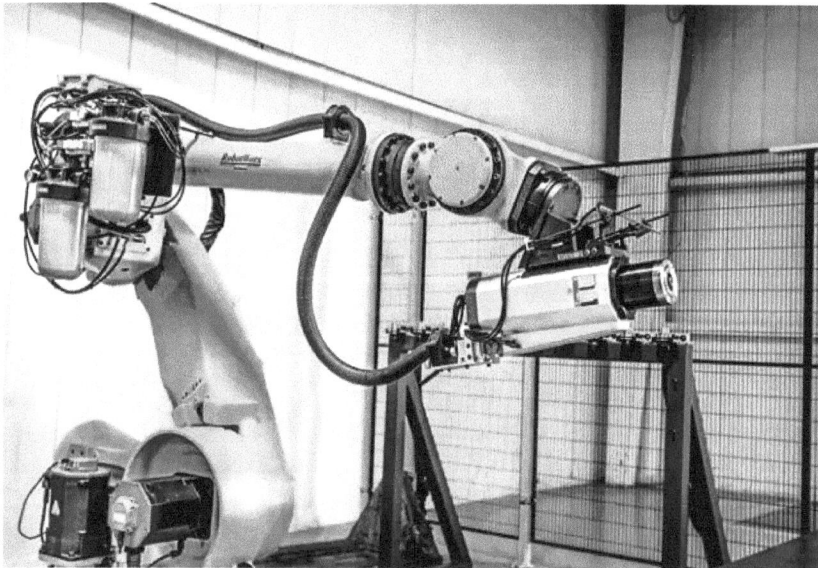

Fig. (3). Robotics in Aviation Industry.

F. Robotics in Defense Sectors

This sector is arguably one of the critical aspects of every land that desires to reinforce its defense system. From Fig. (**4**), we can see that robots help you get close to inaccessible and dangerous zones during the war [9]. They help ensure the safety of soldiers and are used by the military in combat scenarios. In addition to combat support, robots are also used for anti-submarine operations, fire support, combat loss control, strike missions, and construction machinery.

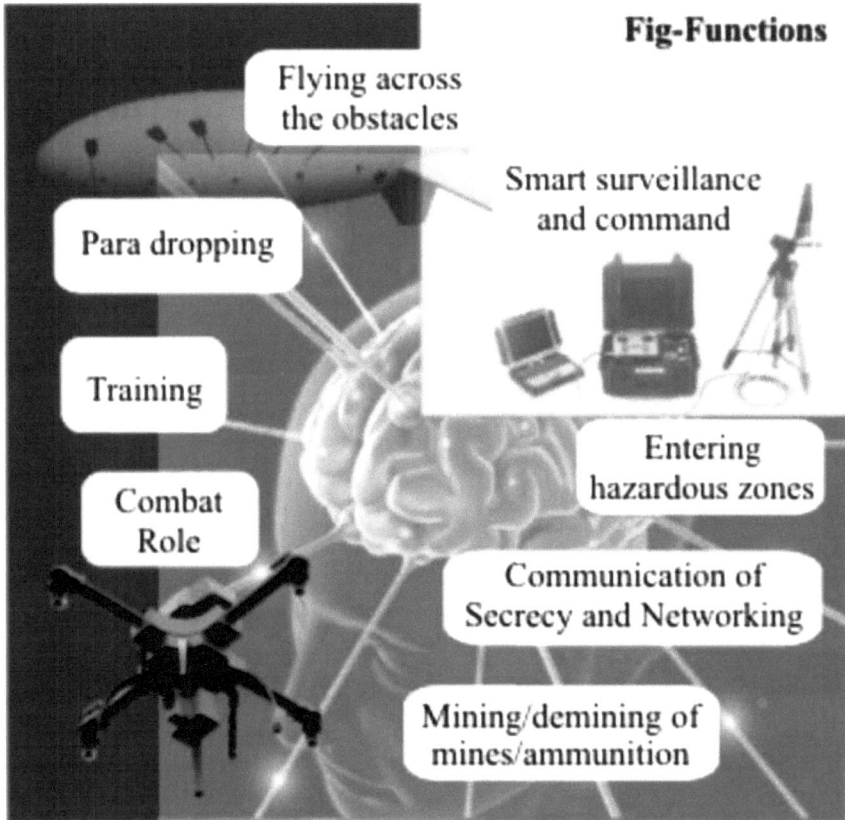

Fig. (4). AI in Military.

G. Robots in the Mining Industry

Robotics are beneficial for various mining applications such as automatic grading, drilling and transportation. Robot mining can guide flooded passageways and detect valuable minerals by operating cameras and additional detectors. In addition, robots assist in drilling to detect gases and other substances to protect people from harm and injury. Robot climbers are used for space exploration, and underwater drones are used for ocean exploration.

LIMITATIONS OF AI

Robots with artificial intelligence and security benefit from artificial intelligence in every aspect of our lives. Developing robots that build machines and computers is complicated and needs more funding to complete an AI project. These Artificial Intelligence-based software programs must be produced regularly to meet the demands of changing surrounding [10]. There is no doubt that machines paint more effectively than people, even though they cannot substitute humans in team leadership. They bond with humans quickly and have no emotions, moral values, or creativity. The biggest drawback of AI is that it only functions for programmed or given instructions. This is not as formulaic as humans and can crash or produce unexpected results. Human dependence on AI technology could pose problems in forthcoming eras. Values and righteousness are intrinsic human traits demanding integration into machines [11]. Building an ethical machine is not easy, as the engine runs fast and must do all the repetitive tasks. We ought to believe in the concessions of AI and the essence of human life that can affect malicious activity.

CONCLUSION

From past successes and failures, there are many lessons learned in artificial intelligence, which we need to understand the specific progress in projects and research ideas. Along with the increase in the impact of technology on society, there are fears that sustained advances in AI require a rational and harmonious interplay between application-specific schemes and innovative investigation opinions. Counting to the unprecedented enthusiasm for Artificial Intelligence is the concern about the technology's impact on our society; we cannot prohibit AI from enhancing our lives in numerous ways. From the blossoming of AI and its possible adverse consequences being mitigated prematurely, we require a straightforward procedure to manage the ethical and legal challenges. Such situations should not inhibit the advancement of AI but rather facilitate the consequence of an organized framework for future AI to flourish. Surpassing all, it is essential to understand science fantasy from its actual reality. With endurable allocation and trustworthy investment, AI will transform the destiny of our civilization: our energies, our living conditions, and our economy.

CONSENT FOR PUBLICATON

I, Pingili Sravya, give my consent for the publication of identifiable details, which can include photograph(s) and/or videos and/or case history and/or details within the text ("Material") to be published in this chapter.

ACKNOWLEDGEMENTS

I thank my professor Dr. Hemachandran and my mentor Dr. Ezendu Ariwa for their expertise and assistance throughout all aspects of our study and for their help in writing the chapter.

REFERENCES

[1]   Tilahun Mihret Estifanos, "Robotics and Artificial Intelligence", Available at: https://www. researchgate.net/publication/344784883_Robotics_and_Artificial_Intelligence(Retrieved on: August 30, 2022).

[2]   Y. Pan, "Heading toward Artificial Intelligence 2.0", *Engineering (Beijing),* vol. 2, no. 4, pp. 409-413, 2016.
[http://dx.doi.org/10.1016/J.ENG.2016.04.018]

[3]   Available at: http://artificialintelligence-notes.blogspot.com/2010/07/perception.html(Retrieved on: August 30, 2022).

[4]   "AI for Advanced Robotics", Available at: https://webinars.sw.siemens.com/en-US/ai-for-advanced-robotics/(Retrieved on: August 30, 2022).

[5]   "The Future of AI: How Artificial Intelligence Will Change the World", Available at: https://builtin.com/artificial-intelligence/artificial-intelligence-future(Retrieved on: August 30, 2022).

[6]   "10.4 Robotics, Artificial Intelligence, and the Workplace of the Future - Business Ethics | OpenStax", Available at: https://openstax.org/books/business-ethics/pages/10-4-robotics-artificial-intelligence-and-the-workplace-of-the-future(Retrieved on: August 30, 2022).

[7]   "Artificial Intelligence for Robotics", Available at: https://www.oreilly.com/library/view/artificial-intelligence-for/9781788835442/(Retrieved on: August 30, 2022).

[8]   "RobotWorx", Available at: https://www.robots.com/articles/robots-in-the-aerospace-industry (Retrieved on: August 30, 2022).

[9]   "Military Artificial Intelligence (Military robots) advantages, disadvantages & applications | Science online", Available at: https://www.facebook.com/Science.online.1/?ref=bookmarks(2019). Available at: https://www.online-sciences.com/robotics/military-artificial-intelligence-military-robots-advantages-disadvantages-applications/(Retrieved on: August 30, 2022).

[10]   M. Chowdhury, and A.W. Sadek, "Advantages and Limitations of Artificial Intelligence", Available at: https://www.researchgate.net/publication/307928959_Advantages_and_Limitations_of_Artificial_ Intelligence(Retrieved on: August 30,2022).

[11]   "Advantages and Disadvantages of Artificial Intelligence", Available at: https://towards data-science.com/advantages-and-disadvantages-of-artificial-intelligence-182a5ef6588c(Retrieved   on: August 30, 2022).

CHAPTER 3

Smart Regime with IoT application using AI

Sri Rama Sai Pavan Kumar[1,*], **Guda Vineeth Reddy**[1], **Sailaja Maggidi**[1] and **Rajesh Kumar K. V.**[1]

[1] School of Business, Woxsen University, Hyderabad, India

Abstract: The Internet of Things (IoT) has made it possible for previously unconnected items, such as vehicle engines, to be connected to the network, leading to the emergence of numerous active data streams. The IoT and big data analytics have made considerable strides, opening up intriguing new possibilities for medical and healthcare solutions. Many organisations still struggle with the usage of AI and ML technology when attempting to expand their digital transformation programmes and utilise IoT data.

The most current trends involve modifying IoT data for smart applications using artificial intelligence techniques. Numerous apps use data science and analytics to extract conclusions from gigabytes of data. However, these applications do not deal with the issue of constantly identifying patterns in IoT data. The introduction of the IoT and the cloud has further enhanced things by offering smart business recommendations as well as insights into how people operate and how lives are changing. We discuss a variety of AI capabilities and how to apply them to IoT devices in Hands-On AI for IoT.

The logic-based substrate provides low energy footprints and higher cognitive accuracy during training and inference, which is a crucial requirement for effective AI with long operating life. The use of AI in the industrial sector has enormous potential. However, it frequently necessitates expensive and resource-intensive machine learning professionals as well as in-depth knowledge of complex statistics and how they are implemented in practical use cases.

Keywords: IoT and AI combined, Tools and innovations, AI for IoT.

INTRODUCTION

In 1999, the phrase "IOT" first surfaced. This was introduced by Kevin Ashton. But at the start of 1990, a toaster was introduced as the first Internet of Things (IoT) gadget [1]. Kevin Ashton's study was crucial to the development of the Internet of Things. He put out the most effective technique for letting computers

---

* **Corresponding author Sri Rama Sai Pavan Kumar:**School of Business, Woxsen University, Hyderabad, India; E-mail: pava.kumar_2023@woxsen.edu.in

Hemachandran K., Raul V. Rodriguez, Umashankar Subramaniam & Valentina Emilia Balas (Eds.)

transmit raw data without human interference [2]. He then introduced sensors and Radio Frequency Identification Devices (RFID) that gather data, are connected to the internet, and instantly feed it to the computer. Since that time, IoT has made it possible for previously disconnected items to be connected to the internet, such as actuators, automobiles, sensors, and smart mobile devices. This has resulted in the development of numerous continuous data streams. IoT applications are currently deployed in industries including agriculture and home appliances [1].

A large amount of information is being generated due to the IoT devices' unprecedently rapid proliferation. Now that sensing technologies have advanced, breakthroughs for the fourth industrial revolution are possible [3]. The majority of the sensitive data collected by IoT devices have been stored using the cloud paradigm, but the system has been rendered ineffective because it takes too long for data to be transferred from cloud data centres to end users. In order to provide services like the movement of money, supply chains, people, and interactive art production, among others [4, 5], IoT devices send a lot of useful data to the network for calculation. Artificial intelligence techniques have lately been used to manipulate IoT data for smart applications. These devices spawn a new generation of AI-powered IoT applications because of the massive amount of data they produce. These applications are designed to make vital decisions in the real world instantly rather than transmitting data to cloud servers. IoT now refers to a wide range of items that are linked to the internet, such as smartphones, actuators, and sensors. The fog network is another common name for it [6]. The great majority of AI systems today are based on the neural network concept. In the end, this makes things simpler and offers people control over the potential of IoT data. Fig. (1) shows two different types of IoT sensors [7, 8].

The analysis of complicated data using algorithms, data mining, and diverse technologies is covered in this chapter along with IoT implementations and approaches for integrating AI into daily life. There has been a strong movement recently to replace arithmetic with binary logic as the basic building blocks. It is laborious, though, to create IoT applications employing AI approaches. The authors offer creative strategies that enable programmers to swiftly incorporate an AI mechanism into existing IoT applications *via* a graphical user interface. IoT apps can easily incorporate AI *via* AI speak. This chapter offers knowledge that doesn't require code to use. Our goal is to assess the learning automata methods applied in the IoT for energy efficiency. We will verify the effectiveness of the AI with the aid of IoT-scale datasets to evaluate either single- or multi-class applications in the ecosystems as well as investigate the impact of noisy inputs on the overall learning efficiency. Three-dimensionality was added to scale it. Fig. (2) in the text below illustrates it in an odd way. These things might include somebody sporting a wristband or even a smartphone, RFID-tagged animals, as well as commonplace appliances like refrigerators, washing machines, and coffee makers [8, 9].

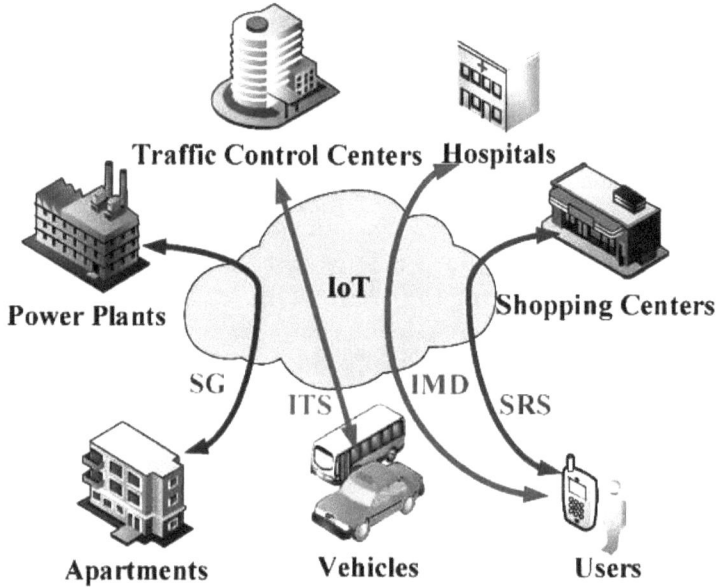

Fig. (1). Types of sensors used in IoT [10].

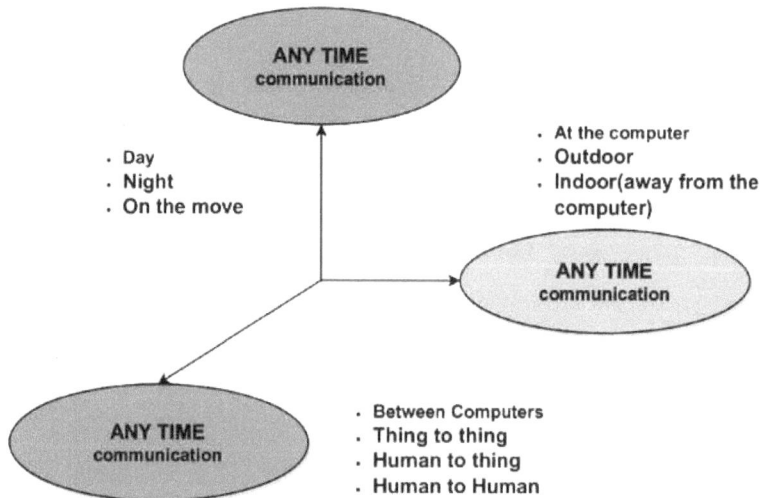

Fig. (2). Dimensions in IoT [11].

These real-world tangible objects can be sensed, automatically triggered, and connected to things that are not immediately apparent but exist as data and may be preserved, examined, and made available. To connect to the internet, these components are required. They might also be able to detect motion, collect data, use cloud computing, and analyse data. How to handle actuators, utilise machine learning, train models, test the models' memory and precision, and link sensors are just a few of the topics discussed in AI talks, taking into account the complexity of the system's real-time operation as well as validation.

Fig. (**3**) shows how an IoT machine-learning process for a smart home application is built using a simple functional block configuration. In this illustration, the home's owner can first be seen utilising a remote control to operate the actuators, such as a fan [12]. Several sensors, including those for temperature, humidity, and other variables Fig. (**3**), produce the data from the sensors in real-time, which is used to automatically run the fan.

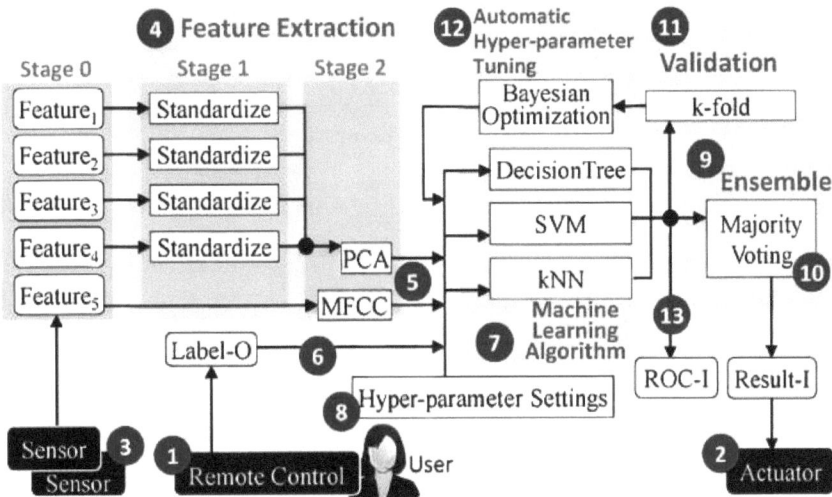

Fig. (3). Applications of machine learning in smart homes.

LAYERS OF IOT

Similar to the OSI model for the internet, Fig. (**4**) shows the six layers that make up the IoT architecture. Four of these layers are horizontal and two are vertical. Management and security are the final two vertical layers, and they are distributed among the four horizontal layers as follows:

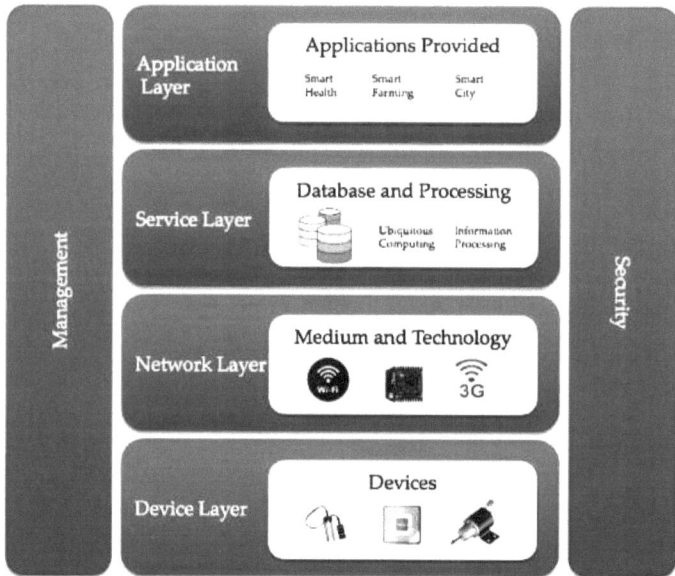

Fig. (4). IoT layers of reference model [15].

1. The bottom layer of the stack is the device layer, also known as the perception layer. This layer stores the physical components required to sense, control, and collect data from the real world. The sensors, RFID, and controllers that make up the perception layer's hardware are:

• The network layer, which is placed on top of wired or wireless networks, aids in data transfer and communication. This layer provides protection for data transfer from the device layer to the information processing system. The networking layer is composed of the technology and the transmission medium. Examples include Bluetooth, Wi-Fi, UMTS, ZigBee, and 3G.

• Service Layer: This layer will always be in charge of overseeing service administration. Through the network layer, it gathers data, stores it in a dataset, analyses it, and then starts to fully automate decisions based on the findings.

• Programmes that solely consume data from the Service Layer are controlled by the Application Layer. IoT can be used to implement a number of applications, including smart cities, precision agriculture, and smart buildings, to mention a few.

PRIMARILY WE ARE USING THESE TECHNOLOGIES IN: WEARABLES

While it is true that at the end of the day, when heart rate is displayed, it displays an average heartbeat, similarly, there are many applications, which can record health conditions. Smart watches also have different functions, including showing heart rate and blood pressure, workout and activity records, *etc*. As seen in Fig. (**5**), consumers will be informed. Due to their unique electrical, mechanical, and chemical characteristics, stretchy and soft materials are crucial for wearable electronics research and development. Currently on the market, wearable electronics are made primarily of metals and semiconductors, which have very little structural flexibility and stretchability [10].

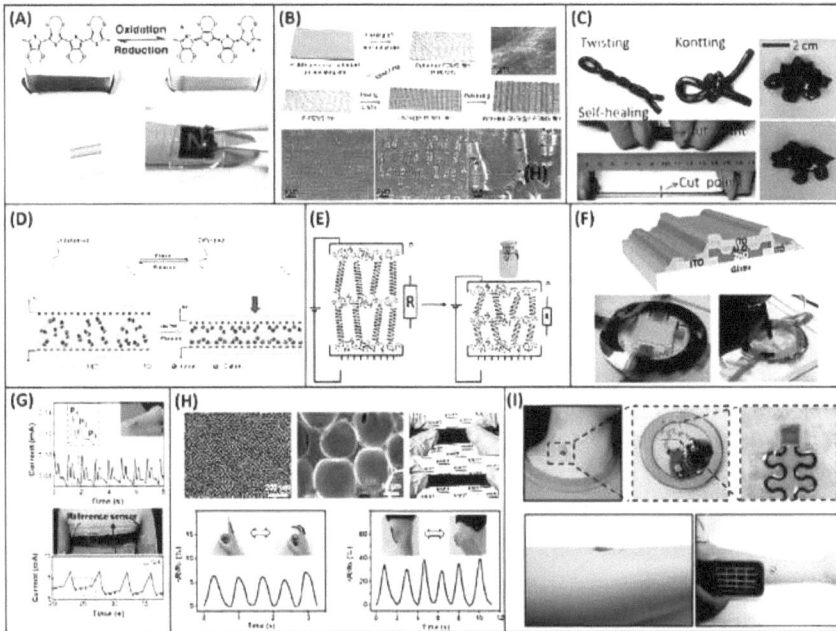

Fig. (5). IoT detection devices [12].

SMART HOMES AND BUILDINGS

Every home has a smart Android television, and as can be seen, some of them are linked to Alexa. Alexa or Google Assistant responds instantaneously and without the use of a remote control when asked to change the channel or switch on the TV. It also works with televisions, air conditioners, and other devices [13]. Smart cities tend to have intelligent lighting systems that, for example, automatically turn on when they sense an approaching person. A sanitizer dispenser, which works similarly and is depicted in Fig. (**7**), was popular in former times and automatically dispenses the liquid when a person places his hand below the dispenser. A robotic Hoover vacuum cleaner that can be controlled by Alexa or Google Assistant and cleans on its own is shown in Fig. (6).

Fig. (6). Smart Vacuum cleaner [14].

Fig. (7). Smart Hand wash [8].

Contrarily, IoT-enabled or smart buildings help people live or work in happier environments while also helping to conserve resources. Additionally, as seen in Fig. (8), more sophisticated sensors have been put in the buildings to track resource usage and proactively detect resident needs. These advanced devices and sensors collect data for monitoring remote buildings, power, safety, landscape, HVAC, lighting, and other areas. This data is then used to forecast operations that, depending on the situation, could be automated in order to maximise efficiency and save time, money, and resources [15].

Fig. (8). Smart Building installations [16].

Self-Driving Vehicles

As seen in Fig. (9), a self-driving automobile is a driverless vehicle that can operate automatically and utilise a range of sensors to identify its environment. The main types of sensors used in self-driving automobiles include thermographic cameras, radar, lidar, sonar, GPS, odometry, and inertial measurement units [7].

From level 0 to level 6, a self-driving automobile primarily has six levels of SAE automation.

Driver assistance at SAE level 0 requires full human assistance because there is no automation at this level.

SAE level 1: driver aid: Driver assistance is provided at this level. There will be very little automation in this, so it relies on human help and the driving assistance system, which controls either the steering or the acceleration.

Partial automation at SAE Level 2: It focuses solely on driver support, and the system will be able to control acceleration and steering.

SAE level 3: conditional automation: At this level, we can observe that the system will primarily use conditional automation to control steering, accelerate, and monitor the driving environment. Only some driving modes will be compatible with automation, therefore, human intervention is required.

SAE level 4: High automation at SAE level 4 means that there will be less need for human assistance when driving and that many driving modes will be automated.

SAE level 5: Full automation at SAE level 5, the highest degree. At this level, there will be complete automation and it will function in all driving modes, so it is self-sufficient.

In self-driving automobiles, we may programme the destination position using a GPS sensor, which will arrive at the destination without any aid from a human, and by utilising a radar sensor, it will detect every item, while the car is travelling, to prevent accidents.

Fig. (9). Vehicle detection.

Security Devices

The majority of the security features we have seen in cars sound an automatic horn if someone tries to open the door, and we have seen door locks that sound an emergency siren if someone tries to open the door without the key.

Traffic Control

AI can be used effectively to optimise traffic lights through object detection algorithms. When any emergency vehicles, like ambulances, are on the road, and if there is heavy traffic, the object will be detected, as shown in Fig. (**10**), and it will give a special route or place to pass the vehicle. The AI-based traffic management system is used to collect and then analyse traffic data in order to provide a solution for traffic in smart cities.

Fig. (10). Vehicle detection (2) [9].

Face, Age & Height Detection System

We have a difficult time estimating the number of persons using high-definition cameras in the specific road transport system. However, we can soon collect the data and determine each person's height, face, and age using Artificial Intelligence approaches. When a certain criminal conduct (which can violate a traffic law) is committed, the credentials can be determined. The majority of self-driving cars are manufactured by Volvo, Mahindra, Mercedes-Benz, Nissan, Tesla, *etc.*

IoT in the Healthcare Industry

Artificial intelligence for providing care for cancer patients diagnosis can be immensely complicated for doctors in making decisions about diagnosing a primary or secondary cancer, so the AI model will help the streamline process by taking information from several sources. Digital health has received a lot of attention recently.

BLOOD TEST   DIAGNOSIS

X RAY IMAGES ────────▶ AI MODEL ────────▶ TREATMENT OPTIONS

GENETIC INFORMATION   PROGNOSIS

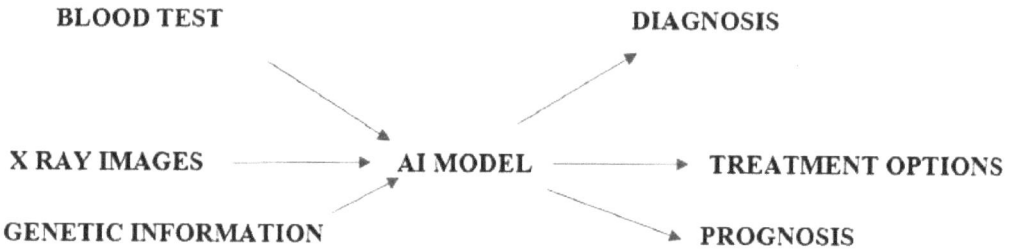

Fig. (11). AI Model.

The trained AI model can quickly consolidate this data, and it provides accurate predictions of the patient diagnosis. A mobile-based diagnostic software, using AI, detects skin diseases, but it does not give perfect diagnosis, so there is a need to implement the successful artificial intelligence in the health care industry. The new regulatory frameworks should be established in consultation with AI developers, health care practitioners, advisors as well as patients to bring the best out of AI to give a perfect diagnosis is shown in Fig. (**11**). Digital health includes overlapping fields, such as artificial intelligence, the IoT, electronic health, telehealth, and big data analysis and implementation. Given the significant room for innovation in the field of digital health, the 'World Health Organization' released a set of recommendations previously this year, and advised prospective researchers and innovators to use this technology to develop data-based interventions in practical settings to enhance patient outcomes. Hence, we highlight a few recent advances in ophthalmology with an emphasis on artificial intelligence and other digital innovation, especially those which are now available in the clinical setting and that could be put into practice shortly.

IoT in the Agriculture Sector

Agricultural efficiency is improved, crop yields are enhanced, and food production costs are reduced because of AI, machine learning, and IoT sensors that generate real-time data for algorithms. The world's population is expected to increase to 2 billion people by the year 2050 (Fig. **12**), where necessities will also be going up by up to 60 percent.

Fig. (12). Smart agriculture [9].

By 2050, an additional 2 billion people are expected to inhabit the planet, and AI and ML have already demonstrated their potential to fill a significant gap by monitoring each crop field's real-time video feed with AI and machine learning-based surveillance systems, among other applications. According to the Economic Research Service of the U.S. Department of Agriculture, the business of growing, processing, and distributing food in the United States alone is worth $1.7 trillion.

Due to the current labour shortage in agriculture, smart tractors, Agri-bots, and robotics powered by AI and machine learning are emerging as a practical solution for many remote agricultural operations. AI can be used to identify patterns in large data sets and determine their orthogonality in real-time, both of which are essential for crop planning. Yield mapping is a farming technique that uses sensors to enhance pest control.

Smart City

For its government and citizens, a smart city may solve problems like traffic, public security, energy management, and more by utilising cutting-edge IoT technologies. It may also have intelligent parking, intelligent mass transit, and other intelligent features. It can maximise the use of the city's infrastructure and the standard of living for its residents.

IOT COMBINATION OF DATA SCIENCE AND AI

Intelligent analysis and understanding of this data are required, which can be accomplished with the least amount of human intervention thanks to the tools and models of AI. IoT provides a lot of data, but only 10% of it is being collected now, and of the 10%, the bulk remains time-dependent and loses its worth in milliseconds. There are several issues with this:

• Preserving the outcomes of current events

• Analytical queries are run over stored events.

• Finding patterns in data using AI, ML, and DL methods to forecast results. For data mining, all industries employ the same technique.

Fig. (**13**) shows the stages of the Artificial Intelligence of Things (AIOT), a recently developed technology that combines AI and IoT. AIOT is intended to build human-machine interactions and improve data management and data analytics.

CAMERA SENSORS　　　　**DEEP LEARNING**　　　　**METADATA**

❖ **INTEGRATIONS**
❖ **NOTIFICATIONS**
❖ **ANALYTICS**

Fig. (13). Stages of AIOT data processing.

Edge AI, also known as edge intelligence, is a concept that enables data analytics and automated decision-making by equipping physical objects with software for sensors and other purposes.

The Data Mining Process in IoT

The cross-industry standardised method for data mining, developed by Chapman *et al.*, is the most well-known data management strategy addressing IoT issues. This process model outlines responsibilities, which must be met to effectively finish DM [16]. These distinct phases make up this vendor-neutral process, as shown in Fig. (**14**).

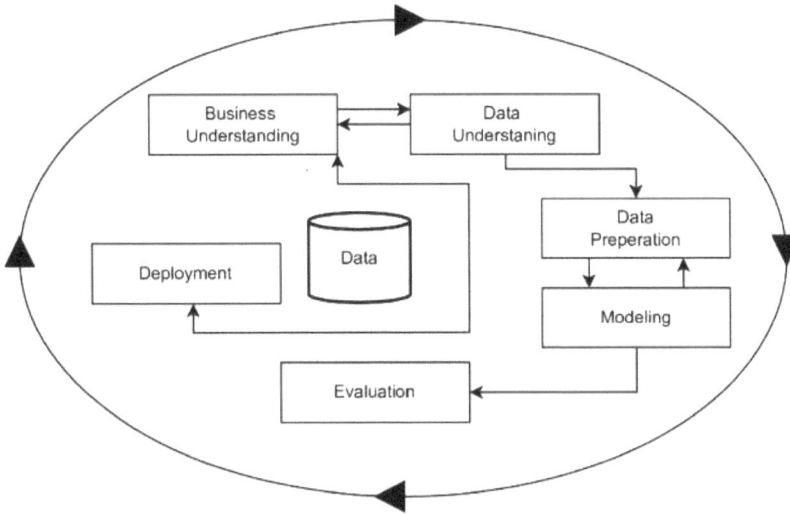

Fig. (14). Process of data mining [15].

1. Knowing what kind of business it is.

2. Recognising the kind of data.

3. Getting the data ready.

4. Data modelling.

5. The changing data and deployment of data.

The many data mining stages are illustrated in the diagram below:

AI LIBRARIES AND THEIR ROLE IN IOT APPLICATIONS

Python will be used during coding in this book due to its popularity in AI and IoT platforms. To do AI/ML analytics on the data, specific supporting libraries like NumPy, pandas, SciPy, Keras, and TensorFlow will be required. We'll be using Seaborn and Matplotlib for visualisation.

The table below covers a wide variety of open-source, free AI tools that may be downloaded from the Internet (see Table **1**), although most of the current IoT applications require extensive programming to integrate AI.

Table 1. AI libraries used in IoT Applications.

Type	Library Name	Commits
Core Libraries	NumPy	15,980
	SciPy	17,213
	Pandas	15,089
Visualisation	Matplotlib	21,754
	Seaborn	1699
	Bokeh	15,724
	Plotly	2486
Machine earning	SciKit-Learn	21,793
	Keras	3519
	TensorFlow	16,785
	Theano	25,870
NLP	NLTK	12,449
	Gensim	2878

Keras

A high-level API named Keres is built on top of TensorFlow. It enables quick and simple prototyping. It supports convolutional neural networks, recurrent neural networks, and a hybrid of the two. It performs well on CPUs and GPUs alike.

TensorFlow

The programme was developed by the Google Brain team and is an open-source framework called TensorFlow that has deep neural network implementation techniques and APIs. This can be used to work on a variety of platforms, including distributed, mobile, GPU, and CPU. Python, R, C++, Java, and Go are all compatible. TensorFlow makes it straightforward to use models in practical contexts and deploy them.

The optimizer in TensorFlow streamlines the process of training deep neural networks by mechanically calculating gradients, which can subsequently be used to modify weights and bias.

In TensorFlow, a programme is divided into two components.

1. The computation graph

2. The execution graph

NumPy

IoT uses NumPy features to read sensor data in bulk from the system's built-in databases. IoT time series data, a tool for scientific computation, makes it easier to build a dataset for testing.

MATPLOTLIB

A tool used for visualising the operations offered by various conversations and graphical representations is called Matplotlib.

INNOVATIVE MILESTONES CAN BE ACHIEVED IN IOT USING AI

The Internet of Things (IoT) is changing how we interact with our gadgets at home, with smart devices at work and in the education sector, with diseases and developing virus-resistant methods in agriculture, as well as with space technology. Every field where there is human interference, it is used [17]. This network-connected technology amasses a vast amount of data regarding our beliefs and behaviours. IoT devices produce 5 quintillion bytes each day, or 1 million gigabytes or 5 exabytes, of data. New discussions have been sparked by driverless cars driven by artificial intelligence and automated supermarkets run by cooperative robots (co-bots) that work without human supervision [15]. Broadband wireless internet connectivity, tiny sensors embedded in both living and non-living objects, such as milk cartons in smart refrigerators and house cats, as well as AI and co-bots that process the Big Data received by sensors, are the foundation of the Internet of Things (IoT). AI's most important IoT-based applications fulfil human needs and improve living conditions.

First, artificial intelligence (AI) and deep learning, a subset of machine learning that has gained appeal in the field of digital health, are driving the digital revolution. In the fields of image, audio, and motion detection, as well as natural language processing, deep learning has been found to outperform conventional feature-based machine learning techniques. Deep learning has been proven to be very successful in the diagnosis of diseases in the field of medicine, such as tuberculosis from chest X-rays, malignant melanoma from skin pictures, and lymph node metastases from breast cancer from histology. The majority of deep learning systems in ophthalmology place a strong emphasis on image recognition to detect diabetic retinopathy, diabetic macular oedema, glaucoma, age-related macular degeneration, retinopathy of prematurity, and cataract using fundus images and optical coherence tomography [13]. IoT can be used in the educational process to improve standardised testing, encourage students through adaptive and personalised learning, automate common question responses, and improve the caliber of online examinations. It frees educators from their conventional roles and empowers them to actively participate in innovation and the personalization of the educational experience [12]. We employ artificial intelligence techniques to spot pupils who are likely to leave the class early and to motivate them to do better. The ambient and embedded wearable sensors in the adaptive education IoT framework assess bio signals with biofeedback, such as heart rate variability or HRV to improve students' self-reflection and make learning more personalised. The preliminary analyses show a strong correlation between the self-reported data and the bio signal measurements.

CONCLUSION

The Internet of Things (IoT) brings together a wide range of network-connected devices to provide intelligent, sophisticated services that help safeguard user privacy from hazards, including surveillance, jamming, denial of service, and IP spoofing. Cybersecurity solutions powered by IoT, cloud, AI, and data analytics are summarised by TechVision Opportunity Engine. These developments help companies defend themselves against threats, data breaches, phishing scams, and other contemporary threats that affect endpoints, the cloud, and other network levels: Automated email security, a Cloud Data Security Framework, AI-based Cyber Risk Modelling and Prediction, Training for Cyber Security Experts to address threats, a Solution for Cybercrime in Software Design through Innovative Cyber Fraud Prevention Methods, assisting companies to lessen the impact of cyberattacks, and safeguarding SMBs from complex threats with a platform for Software Defined Security on IoT Devices for Real-Time Vulnerability Management Security. We found that many IoT devices make analysis and prediction possible, and these capabilities may be supported by big data and AI. In this chapter, a number of IoT platforms and certain well-known verticals were briefly introduced. Additionally, we studied the two distinct deep-learning software packages, Keras and TensorFlow [13]. The introduction of several of the datasets used in the book was the final step. Significant progress has been made thanks to artificial intelligence, and creative solutions have appeared across all sectors of the economy [8]. These solutions have boosted development, enhanced quality of life, protected the environment, and stimulated innovation across all sectors of the economy.

CONSENT FOR PUBLICATON

I, Sri Rama Sai Pavan Kumar H, offer my permission for the publication of personally identifiable information, including photographs, videos, case histories, and/or textual details ("Material"), in this chapter.

ACKNOWLEDGEMENTS

I am grateful to my lecturer, Dr. Rajesh Kumar KV, for his knowledge and assistance during every stage of our research as well as for his assistance with the chapter's writing.

REFERENCES

[1] H.M. Gomes, and J.P. Barddal, "Enembreck a23-gomes-apndx.pdf. (n.d.)., from Google Docs website", Available from: https://drive.google.com/file/d/1nphsIP7haR8yVpsGUjtrTBYaaXm1BnNQ/ view (Retrieved September 1, 2022).

[2] "My Drive. (n.d.). Retrieved September 3, 2022, from Google Drive website", Available from: https://drive.google.com/drive/my-drive

[3] "Innovations in IoT, Machine Learning, and Artificial Intelligence-Based Security Solutions 2017 - Research and Markets. (2017, June 19). Retrieved September 1, 2022, from Business Wire website", Available from: https://www.businesswire.com/news/home/20170619005795/en/Innovations-in-I-T-Machine-Learning-and-Artificial-Intelligence-Based-Security-Solutions-2017---Research-andmar-kets.

[4] A. Wheeldon, R. Shafik, T. Rahman, J. Lei, A. Yakovlev, and O.C. Granmo, "Learning automata based energy-efficient AI hardware design for IoT applications", *Philos. Trans.- Royal Soc., Math. Phys. Eng. Sci.,* vol. 378, no. 2182, p. 20190593, 2020.
[http://dx.doi.org/10.1098/rsta.2019.0593] [PMID: 32921236]

[5] N. Zafar, "(2021, December 21). Top 8 AI Applications in Agriculture. Retrieved September 1, 2022, from Revolveai website", Available from: https://revolveai.com/ai-applications-in-agriculture/ #:~:text=Applications%20of%20AI%20in%20Agriculture%201%201.%20Precision,...%208%208.%2 0Mapping%20Yield%20with%20Demand%20%209.https://en.wikipedia.org/wiki/Self-driving_car

[6] "Future Internet: The IoT Architecture, Possible Applications and Key Challenges. (n.d.). Retrieved September 1, 2022, from IEEE Xplore website", Available from: https://ieeexplore.ieee.org/ abstract/document/6424332

[7] "Role of IoT in road safety and traffic management. (n.d.). Retrieved September 1, 2022", Available from: https://indiaai.gov.in/article/role-of-IoT-in-road-safety-and-traffic-management

[8] "The Fourth Industrial Revolution . (n.d.). Retrieved September 1, 2022, from Google Books website: ", Available from: https://www.google.co.in/books/edition/_/ST_FDAAAQBAJ?hl=en

[9] S. Lucero, n.d., Available from: https://cdn.ihs.com/www/pdf/enabling-IOT.pdf

[10] K. Guo, Y. Lu, H. Gao, and R. Cao, "Artificial intelligence-based semantic internet of things in a user-centric smart city", *Sensors,* vol. 18, no. 5, p. 1341, 2018.
[http://dx.doi.org/10.3390/s18051341] [PMID: 29701679]

[11] "Hands-On Artificial Intelligence for IoT", Available from: https://www.packtpub.com/product/ hands-on-artificial-intelligence-for-IoT/9781788836067

[12] Y.W. Lin, Y.B. Lin, and C-Y. Liu, "AItalk: A tutorial to implement AI as IoT devices", *IET Netw.,* vol. 8, no. 3, pp. 195-202, 2019.

[http://dx.doi.org/10.1049/iet-net.2018.5182]

[13]   "IoT Standards & Protocols Guide. (n.d.). Retrieved September 1, 2022, from Postscapes website", Available from: https://www.postscapes.com/internet-of-things-protocols/

[14]   "velway IoT vacuum cleaner. (n.d.). Retrieved September 26, 2022, from Google Search website", Available from: https://www.google.com/search?q=velway+IoT+vacuum+cleaner&tbm=isch&ved= 2ahUKEwiq2c7bzrL6AhWnzaACHW3cD2MQ2cCegQIABAA&oq=velway+IoT+vacuum+cleaner& gs_lcp=CgNpbWcQAzoECCMQJ1DzFViMJmCpL2gAcAB4AIAB8gOIAY4XkgEFMy03LjGYAQC gAQGqAQtnd3Mtd2l6LWltZ8ABAQ&sclient=img&ei=-q8xY6r0DKebg8UP7bi_mAY&bih=656& biw=1536&rlz=1C1CHBF_enIN1000IN1000#imgrc=zxCaZU8MhzCcSM

[15]   O. Marbán, G. Mariscal, and J. Segovia, "(2009, January 1). A Data Mining & Knowledge Discovery Process Model. Retrieved September 1, 2022, from unknown website", Available from: https://www.researchgate.net/publication/221787522_A_Data_Mining_Knowledge_Discovery_Proces s_Model

[16]   Machine Learning Techniques in IoT Applications, "A State of The Art", Available from: https://www.taylorfrancis.com/chapters/edit/10.1201/9781003124252-6/machine-learning-techniques-IoT-applications-state-art-shaw-laxmi-narayan-sahoo-rudra-hemachandran-kumar-nanda-santosh? context=ubx&refId=8f4a1e66-bb2e-4060-aaf9-83927b68ec30

[17]   "IoT in Industries: A Survey. (n.d.). Retrieved September 1, 2022, from IEEE Xplore website", Available from: https://ieeexplore.ieee.org/document/6714496/authors#authors

CHAPTER 4

Artificial Intelligence in Marketing and Operations

Gaddam Venkat Shobika[1,*], Sourav Chakraborty[1], Varadharaja Krishna[1], Dibya Nandan Mishra[1] and **Pranay Kumar[2]**

[1] *Woxsen School of Business, Woxsen University, Kamkole, Sadasivpet, Telangana, India*

[2] *University of Maryland, Baltimore County, United States*

Abstract: In the last 20 years, artificial intelligence (AI) and machine learning (ML) applications have advanced at an unmatched rate. The development of robotics and automation has been driven by AI technology, and this has substantial effects on practically every facet of the business, particularly supply chain operations. Smart technologies allowing real-time automatic data collection, analysis, and prediction have been widely incorporated into supply chains. We examine the current uses of AI in marketing and operations management (OM) and supply chain management in this study (SCM). Since these three industries combined account for the majority of business-related AI advancements as well as expanding problem domains, we focus specifically on innovations in healthcare, manufacturing, and retail operations. We go over the main obstacles and potential uses of AI in those sectors. We also talk about current research.

Keywords: Artificial Intelligence, Marketing, Market Basket Analysis, Operations.

INTRODUCTION

The idea of Artificial Intelligence is primarily based on information technology. It is commonly used with concepts such as robotics and automation. Artificial Intelligence may also be referred to as an algorithm solicitation. The technology of AI is capable of making analytical or cognitive functions. Such may include the capacity for problem-solving as well as learning simultaneously. Mainly there are two divisions of Artificial Intelligence. One being Artificial Narrow Intelligence and the other Artificial General Intelligence. Artificial Intelligence is nothing but a clone of human intelligence that is processed by computer systems. Its applications consist of machine learning, natural language processing and deep learning.

---

[*] **Corresponding author Gaddam Venkat Shobika:** Woxsen School of Business, Woxsen University, Kamkole, Sadasivpet, Telangana, India; E-mail: shobika.gaddam_2023@woxsen.edu.in

Machine learning has contributed to a huge extent in the area of Artificial Intelligence. It is a science that enables a computer to perform based on available datasets even without the requirement of programming. Machine learning has several uses that include data analysis, pattern recognition, predictive analysis and statistical modelling. Deep learning is mainly a subgroup of machine learning, but it works on learning algorithms which do not even require to be managed manually. This comprises the benefit of cloud computing as well as big data. Deep learning has surely helped in the advancement of Artificial Intelligence. Natural language processing aims at speech recognition. It is one of the most essential applications of machine learning and deep learning.

There have been numerous advancements in these areas of technology which have helped in the evolution of Artificial Intelligence in the sections of autonomous robots and vehicles, decision making and image as well as voice recognition. For example, nowadays, we find Siri and Google Assistant on smartphones. This is an application of voice recognition. Decision-making system is found in the areas of education like the IBM Elements. Image detection has made it easier for payment approvals.

In the last 20 years, artificial intelligence (AI) and machine learning (ML) applications have advanced at an unmatched rate. The development of robotics and automation has been driven by AI technology, and this has substantial effects on practically every facet of the business, mainly supply chain operations. Smart technologies which permit real-time automatic data collecting and analysis have been widely incorporated into supply chains. We examine the current uses of Artificial Intelligence in operations management as well as supply chain management in this study. Since these three industries account for most business-related AI advancements as well as emerging problem areas, we focus specifically on innovations in healthcare, manufacturing, and retail operations. We go over the main obstacles and potential uses of AI in those sectors. We also talk about fads.

Application of Artificial Intelligence in Marketing

Artificial Intelligence has made a huge impact in the field of marketing, and there has been research done by various authors in this area over some time. Some of such findings and areas of work are listed below:

The application challenges of self-governing customer experience management are popularly known as CEM, Customer Experience Management [1]. This paper also demonstrated how intelligence networks, as well as critical business value operators, are built with the help of machine learning and artificial intelligence. Artificial Intelligence transformed traditional stores into mart retail outlets.

They demonstrated the Artificial Intelligence supported machine that could track the five senses of humans [2]. It has shown significant results in better consumer-brand alliance in e-commerce. Studies based on enhancing the cement of customer experience by using Artificial Intelligence had relevant findings regarding customer experience augmented through AI-driven bots [3]. Apart from segmentation, targeting and positioning, Artificial Intelligence also helps the merchants or dealers in visiting only the state orientation of the company [4].

Artificial Intelligence has a significant impact on the Marketing Mix, specifically in creating as well as developing innovations in merchandise optimization, sales management and delivering customer service. Some practical applications of AI in the field of marketing are listed below:

1. Text processing technology - The use of a virtual assistant in shopping complexes [Alpine Artificial Intelligence]. A navigation system (GPS) not only shows the correct route to a particular place but also suggests the attractions found on that route [Naver]. The alterations in the customer service methods and analysis of the statements made by the insurance companies as well as telecoms to identify the unhealthy situations [Touchpoint]. The handling of client requests through the virtual assistant implanted in a cell phone bank application. A virtual assistant helps not only in the purchase of bank products but also provides essential information such as the location of cash machines and bank branches [ING Bank Slaski].

2. Voice Processing Technology - Virtual assistants in today's world help in task execution like Google Assistant and Siri. Amazon Alexa is another example of voice processing technology that has helped in the area of marketing.

3. Decision Making - The system of brand-new product suggestions in the case of Netflix and Amazon. The appearance of sun subjection time and location is based on the user's cell phone data. The production of a custom savings plan that matches an individual's financial capabilities through an application [Plum]. Based on data analysis, Artificial Intelligence helps to come up with specific suggestions regarding the campaign plan [Harley Davidson and Albert AI]. This is the process of contemporizing the user data from all the points of contact with a particular brand, such as email, websites as well as social media. This application of Artificial Intelligence can be found in Adidas and Salesforce, which have significantly shown better customer service over time.

4. Autonomous robots and vehicles - Autonomous shop which provides fresh goods and magazines. The shop was verified in Shanghai [Moby Mart]. Another example will include the service-free shops in the case of Amazon Go and Alibaba. The system to understand and forecast the next movements of

cyclists, pedestrians as well as cars in Tesla is another brilliant application of Artificial Intelligence in this case.

5. Processing and Image Recognition Technology - In Adobe Sensei, embedded machine learning helps to automatically frame images based on the requirements of communication channels as well as the brand. Face recognition is a process to make payments in KFC. There are several other examples as well that show the image recognition technology of artificial intelligence.

IMPACT OF ARTIFICIAL INTELLIGENCE ON CUSTOMERS

As we already know, that Internet did bring a revolution as well as several benefits from a customer's viewpoint, like specific product suggestions, customer service specialty and reduced purchase time. Artificial Intelligence has taken this a step further and offers the latest opportunities in the marketing scheme. Some of the impacts that AI offers to customers are listed below:

- Faster and more suitable purchasing time. For example, the 24/7 customer service, the improved quality of search engines and the automated payments.
- The new customer experience through the extensive hyper-actualization, and sales facility has impacted to create an additional worth.
- Elimination of the method of category learning has led to a new dimension of a customer- brand bonding.

IMPACT OF ARTIFICIAL INTELLIGENCE ON MARKETING

1. Removal of laborious as well as time taking projects. The repeatable and routine activities are automated by Artificial Intelligence. Example - Processing, Data analysis and collection and image search.
2. Artificial Intelligence has helped in redefining the method by which the value is transferred to the customer and thereby raising the role of getting brand new solutions by design innovations.
3. Building brand new competencies in the marketing team. Artificial Intelligence needs to integrate data scientist skillset and knowledge about the new technology prospects in the marketing team.
4. Complexity of artificial intelligence rises the role of business firms in constructing Artificial Intelligence solutions. This has aided in building a new marketing ecosystem.

APPLICATION OF ARTIFICIAL INTELLIGENCE IN OPERATIONS

Rather than the normal insight shown by creatures, including people, man-made brainpower (AI) is knowledge exhibited by robots. The investigation of canny specialists, or any framework that comprehends climate and acts in a manner that boosts its possibilities of succeeding, has been characterized as the focal point of AI research.

Robots that copy and show "human" mental capacities connected with the human psyche, for example, "learning" and "critical thinking," were recently alluded to as having "man-made consciousness." Major AI scientists have, as of late, exposed this hypothesis and are depicting AI as far as reason and a legitimate way of behaving, which doesn't put limitations on the presentation of knowledge.

A couple of instances of AI applications are state-of-the-art web search tools like Google, suggestion frameworks like YouTube, Amazon, and Netflix, discourse acknowledgement programming like Siri and Alexa, self-driving vehicles like Tesla, mechanized direction, and overwhelming the best essential game frameworks (like Chess and Go). The AI impact is a peculiarity where activities once remembered to require "insight" are presently habitually prohibited from the meaning of AI as machines develop increasingly proficient. For example, although being a typical strategy, optical person acknowledgement is regularly avoided in the classification of things accepted to be man-made reasoning.

Since its foundation as a field of concentrate in 1956, man-made reasoning has gone through various rushes of positive thinking, trailed by difficulties and a decrease in support (known as a "Man-made intelligence winter"), then new methodologies, accomplishments, and expanded speculation. Since its beginning, AI research has tried different things with and deserted many philosophies, including demonstrating human critical thinking, formal rationale, broad information bases, and creature conduct impersonation. AI that is vigorously situated in arithmetic and measurements has overwhelmed the subject in the initial twenty years of the twenty-first 100 years. This approach has been exceptionally compelling in tackling numerous troublesome issues in both industry and the scholarly world.

The numerous subfields of AI research are fixated on objectives and the utilization of strategies. A portion of the standardized objectives of AI research integrates thinking, information portrayal, arranging, learning, normal language handling, detecting, and the ability to move and control objects. One of the drawn-out targets of the area is general knowledge, or the ability to take care of any issue. Man-made brainpower (AI) scientists have incorporated and changed an extensive variety of critical thinking methods, including formal rationale, counterfeit brain organizations, search, and numerical enhancement, as well as approaches from insights, likelihood, and financial matters, to resolve these issues. Software engineering, brain science, semantics, theory, and numerous different disciplines are additionally affected by AI.

The possibility that human keenness "can be so completely portrayed that a machine might be developed to mirror it" filled in as the establishment for the review. As issues of fantasy, writing, and reasoning since artefact, this produced conversations about the psyche and the moral ramifications of making wise counterfeit creatures. From that point forward, PC researchers and thinkers have asserted that computerized reasoning may ultimately address an existential danger to humankind on the off chance that innovation is not utilized for useful purposes.

To guarantee that the items are per client assumptions and to accomplish objectives associated with the framework upper hand, store network the board (SCM) plans to digitalize business processes and coordinate different partners and resources [5]. Due to the exclusive idea of the inventory network, rapidly moving client attention, amorphous choice issues, and the ceaselessly moving status of business processes, these divided arrangements are, nonetheless, not "smart" enough (*i.e.,* not ready to act reasonably founded on the climate) and are not truly suitable for current SCM. It is pivotal to work with the greatest amount of proficiency in every critical activity and business stream in the production network to foster canny, fast, and powerful corporate reaction frameworks.

Man-made brainpower (AI) portrays a machine's ability to get information using involvement and settle on various undertakings with knowledge [6]. For the late headways in profound brain organizations, convolutional brain organizations, numerical enhancement strategies utilized in activities research, limitation programming, and various mathematical techniques, man-made reasoning (AI) is an arising theme in software engineering.

These improvements have permitted PCs to perform things that were already an option exclusively for people. To build sane creatures who can notice and act so that some goal capability is enhanced is the objective of man-made brainpower [7]. The three key mechanical powers can be summarized as rising handling power, rising information volumes, and rising algorithmic complexity.

The cloud's foundation is continuously created. In cutting-edge specialized frameworks, it is the prevalent figuring asset. Various organizations, like Google, Amazon, Microsoft, and Salesforce, are offering dependable PC frameworks through the cloud [8]. With this arrangement, AI can be leased or bought depending on the situation.

Information is accumulated from activities, business exchanges, and sensor inputs as it keeps on expanding. Albeit this information is a significant resource for organizations, they likewise give a huge snag to getting the required information. Calculations utilizing huge information heuristics are the solution to this issue. As well as giving exact data considering a savvy, specific pursuit of the whole arrangement of gigantic information, they can be utilized to get pivotal bits of knowledge into activities and the store network [9].

In numerous applications where individuals used to succeed, calculations (model) are more refined and predominant [10]. This power is a consequence of the past two powers meeting up (cloud framework and information sum). Memory-based separating calculations make up the customary calculations. They have been displaced by more powerful and dependable AI frameworks [10].

AI methodologies offer richer and more adaptable representations of real-world issues, providing effective constraint-based reasoning mechanisms as well as mixed initiative frameworks that enable the involvement of human knowledge. The difficulty is in generating representations that are expressive enough to depict problems in the real world while also ensuring effective and quick solutions.

APPLICATION OF MARKET BASKET ANALYSIS

As digital marketing and analytics keep on flourishing harmoniously, strategically pitching and up-selling have turned into the mantra for the past 10 years. A vital part of inferring buyer experiences is market basket analysis or MBA. In this blog, we provide you with the aftereffect of market basket analysis and how it tends to be used to comprehend the clients better. We likewise investigate some true Market Basket Analysis models and what they have meant for various areas.

The most dependable definition is that it is a data mining procedure that is operated to uncover buying strategies in any retail background. The analysis expects to comprehend purchaser conduct and behavior buying connections among the individuals' purchases. For instance, individuals who purchase green tea are likewise prone to purchase honey. In this way, Market Basket Analysis would quantitatively lay out that there is a connection between Green Tea and Honey. The equivalent goes for bread, margarine, and jam.

This is a method that hunts for blends of objects in buys. Above all, its principle is to understand that customers' buying intention to understand the purchase of a specific item is pretty much prone to purchase a related item.

Architecture

The target of the Market Basket Analysis model shown in Fig. (1) is to distinguish the following item that could create enthusiasm in a client. Thus, advertising and outreach groups can foster more compelling valuing, item situation, strategic pitch and up-sell techniques.

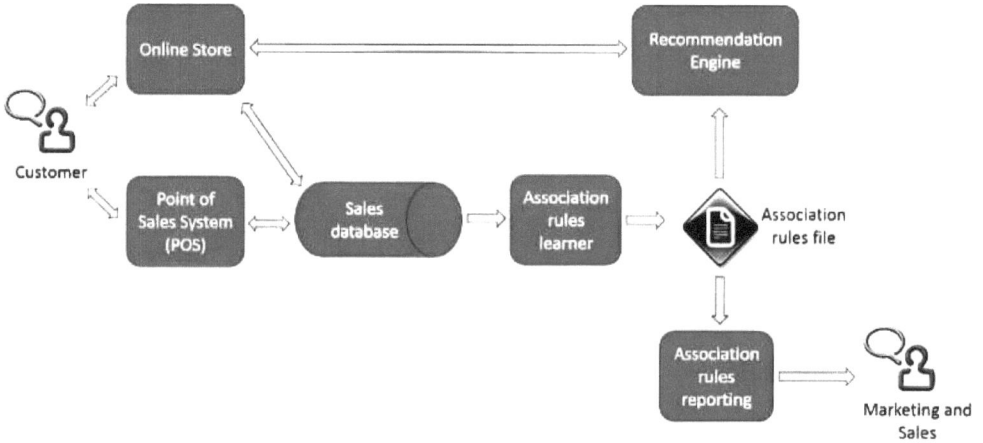

Fig. (1). Architecture of the analysis.

It can assist with foreseeing item deals in unambiguous areas, further developing delivery times and distribution center activities.

Thus, it converts into expanded incomes, bringing down costs and higher benefits.

Market Basket Analysis is modelled on Association rule mining, *i.e.,* the IF {}, THEN {} construct.

Do we figure that the customer will buy the recommended products? Fig. (2) .

Frequently bought together

Total price: $736.97

Add all three to Cart

Add all three to List

i   These items are shipped from and sold by different sellers. Show details

☑ **This item:**Acer Spin 3 Convertible Laptop, 14 inches Full HD IPS Touch, 8th Gen Intel Core i7-8565U, 16GB DDR4... $709.99

☑ NIDOO 14 Inch Laptop Sleeve Water-Resistant Computer Case Portable Carrying Bag for 14" Notebook... $15.99

☑ VicTsing MM057 2.4G Wireless Portable Mobile Mouse Optical Mice with USB Receiver, 5 Adjustable DPI... $10.99

Fig. (2). Example of Amazon's recommendation.

This Amazon feature is fueled by programming that gets a handle on the huge measures of information they gather.

For the client, this is advantageous, as the person in question won't need to return later to purchase another thing. Likewise, for the business, this implies more deals. Everyone wins.

BASICS OF MARKET BASKET ANALYSIS

A regular objective of Market Basket Analysis is to give a bunch of affiliation rules in the accompanying structure:

IF [antecedent] THEN [consequent]

Fig. (3). Formula of antecedent and consequent.

In Fig. (3), the initial segment of the standard is the "body" or "antecedent," while the second piece of the standard is the "head" or "consequent". Moreover, the antecedent and consequent can incorporate many circumstances, making more muddled rules. Example is shown in Fig. (4):

IF ['Smartphone'] THEN ['Case']

A more complex rule could be:

IF ['Smartphone', 'Case'] THEN ['Screen Protector', 'Replacement Battery']

Fig. (4). Formula applied in the example.

Association rules have three reliability measures: Support, Confidence, and Lift.

Support

It measures the frequency of the association rule in the data. For instance, see Fig. (5):

The number of transactions with 'Smartphone,' 'Case,' 'Screen Protector,' and 'Replacement Battery.'/A total number of transactions.

Fig. (5). Support Formula.

A 3% support would imply that 3 out of 100 deals incorporate the 4 referenced things.

Confidence

Confidence estimates areas of strength for how affiliation rule is. Fig. (**6**) shows following is an illustration of how to work it out:

> Confidence = # Transactions with 'Smartphone,' 'Case,' 'Screen Protector, and 'Replacement Battery.'/# Transactions with 'Smartphone' and 'Case.'

Fig. (6). Confidence Formula.

A 40% confidence implies that 4 out of 10 deal exchanges with a Smartphone and a Case likewise incorporate a Screen Protector and Replacement Battery.

All in all, if you purchase a Smartphone and a Case, you are 40% bound to purchase a Screen Protector and Replacement Battery.

Market Basket Analysis Lift

Lift is the proportion of certainty to anticipated certainty.

Expected certainty is the certainty of the "consequent" condition. In other words, the quantity of buys with the resulting condition is partitioned by the absolute number of exchanges. Thus, lift let us know how much better a standard is at foreseeing the result than simply expecting it in any case.

Fig. (**7**) is an illustration of the way to ascertain lift:

> Confidence/(# Transactions with 'Smartphone', 'Case' / # Total Transactions)

Fig. (7). Confidence Formula.

Association Rules Algorithms

The most well-known algorithm producing these guidelines is the Apriori algorithm.

The Apriori algorithm is intended to work on data sets containing exchanges. For instance, things purchased by clients.

First and foremost, it recognizes normal individual components in the dataset. Furthermore, it broadens the set by the accumulation of an ever-increasing extent, if they show up frequently enough. A short time later, it computes the "Support" and "Confidence" of each standard, making it conceivable to look at them as indicated by their significance. Many software design dialects incorporate the Apriori algorithm as a bundle. It is useful to set edges for help and certainty while implementing the Apriori calculation. There are different calculations for producing affiliation rules, like Eclat and FP-Growth. They likewise are great at mining regular itemset tracked down in data sets.

When you run the calculation and create the affiliation governs, the outcome seems to be this, as shown in Table **1**:

Table 1. Output.

Antecedent	Consequent	Support	Confidence	Lift
smartphone	armor case	0,003411	0,46	5,75
	armor case, screen protector	0,001612	0,24	3,00
smartphone, armor case	screen protector, extra battery	0,000157	0,73	214,01
smart WIFI plug	smart WIFI LED light bulb	0,000356	0,58	47,15
	WIFI thermostat	0,000182	0,12	9,76
office laptop	laptop bag	0,001401	0,71	17,75
	laptop bag, wireless mouse	0,000246	0,27	6,75
student backpack	lunch bag	0,002539	0,29	14,50
PC operating system software	Office software, antivirus software	0,001946	0,31	10,33

MARKET BASKET ANALYSIS APPLICATIONS

This part investigates Market Basket Analysis, MBA, models by market portion:

1. Retail: Maybe the most notable MBA contextual analysis is Amazon.com. Any time you view an item on Amazon, the item page consequently suggests, "Things purchased together habitually." It is maybe the least difficult and clean illustration of strategically pitching procedures utilizing an MBA.

Aside from online business designs, BA is likewise generally material to the in-store retail fragment. Supermarkets consider item situation-based and racking advancement. For instance, you are quite often prone to find cleanser and conditioner put exceptionally near one another at the supermarket.

Walmart's scandalous lager and diapers affiliation account is likewise an illustration of Market Basket Analysis.

1. Telecom: With the always expanding contest in the telecom area, organizations are giving close consideration to the administrations that clients are often utilizing. For instance, Telecom has now begun to package TV and Internet bundles separated from other limited web-based administrations to diminish stir.
2. IBFS: Tracing charge card history is an immensely beneficial MBA opportunity for IBFS associations. For instance, Citibank, as often as possible, utilizes deals faculty at large malls to draw likely clients with appealing limits in a hurry. They likewise partner with applications like Swiggy and Zomato to show clients a huge number of offers they can profit themselves using buying through Visas. IBFS associations additionally use container examination to decide on deceitful cases [11].
3. Medicine: In the clinical field, container examination is utilized to decide comorbid conditions and side effect investigation. It can likewise assist with distinguishing which qualities or characteristics are inherited and which are related to neighbourhood natural impacts. DRDO has run an itemized concentrate on related clinical boundaries with the conclusion of mind growth.
4. Operations: Advancing the Supply Chain with Market Basket Analysis. Basket analysis additionally stretches out to the store network. Consider a distribution centre like numerous the nation over lines of racks containing ranges of painstakingly stamped products. To meet customer assumptions, the right merchandise should be picked from the rack, pressed into bundling, and delivered to the client. A similar pick-pack-transport rule applies in a dispersed model when you cannot stockroom inventory.

Assuming things are purchased together regularly, sort out your stockroom and your racks to lessen the time expected to accumulate things for orders. A similar guideline applies to the creation of discrete items: if you have 10,000 SKUs for metal balls, affiliation calculations can assist you with estimating requests.

BENEFITS OF MARKET BASKET ANALYSIS

Despite being a three-decade-old method, market analysis definition stays an important answer for experiences in both the blocks and cement and eCommerce areas.

Expanding market share: Once an organization hits top development, it becomes testing to decide on better approaches for expanding its piece of the pie. Market Basket Analysis can be utilized to assemble segment and improvement information to decide the area of new stores or geo-designated promotions. For instance, if you've at any point considered how there's a McDonald's wherever you go, the response can probably come from MBA.

Behaviour analysis: Understanding the client ways of behaving is a basic stone in the underpinnings of promotion. MBA can be utilized in any place from a basic inventory plan to UI/UX.

Improvement of In-store activities: MBA isn't just useful in figuring out what goes on the racks yet in addition behind the store. Geological examples assume a key part in deciding the prominence or strength of specific items, and subsequently, MBA has been progressively used to improve stock for each store or distribution centre.

Campaigns and Promotions: Not just is MBA used to figure out which items go together, yet in addition, about which items structure cornerstones in their product offering. For instance, organizations might see that habitually restocking connoisseur bread expands the acquisition of other related connoisseur sticks and jams.

Besides, a business can go past appearance-related things to clients. For example: Offer a markdown for additional things at the hour of the offer. Afterwards, send the client a pamphlet or email crusade with appealing item packages. After completing the deal, offer a coupon captivating the client to return and purchase the additional things. On the off chance that you have a physical store, you could put the things near each other.

Recommendations: OTT stages like Netflix and Amazon Prime advantage from MBA by understanding what sort of motion pictures individuals will generally observe habitually. For instance, an individual who evaluated Money Heist exceptionally could likewise be keen on other horror series.

Market Basket Analysis creates a rundown of item sets with their likelihood. Such a rundown can be useful to:

- Know the fascination or repellence between items. Thus, a business can come to better-educated conclusions about item position.
- Rank each arrangement of items by their likelihood. In this way, you can realize which affiliations are predominant over others.
- Figure out which items to put near each other to increment group buys.
- Plan showcasing advancements around items that drive the offer of different items.

- Try not to remember the two items for a similar advancement (Since you realize one will drive the offer of the other).
- Increment strategically pitching and up-selling open doors. You can do this since you have at least some ideas about which items are top merchants and which ones drive additional things' deals.
- Assists in promoting and dealing associations with to appropriate their assets better (Channel advancement).
- Convey internet advertising efforts to clients considering their buying conduct.
- To achieve this, organizations use proposal motors. These are programming parts that make proposal rules accessible to different frameworks. For instance, an internet-based store could incorporate a proposal motor to make offers or item situations.
- To put it plainly, Market Basket Analysis is an important method for utilizing data that you as of now, need to anticipate what clients will purchase.

CONCLUSION

Hence, Artificial Intelligence in marketing is a process of using intelligence technologies to accumulate customer insights, and data and make automated resolutions which would impact marketing attempts. Artificial Intelligence makes the methods of marketing much faster and can boost the yield of investment in marketing [12].

The supply chain (SC) is fundamental for moving merchandise across significant distances and for cultivating associations among different partners, including makers, retailers, wholesalers, planned operations suppliers, and clients. The capacity to make these associations appropriately, quickly, and moderately is the consequence of an effective and productive SC. Data sharing, process incorporation, and joint effort are essential SC achievement components [13]. Therefore, SC should become advanced and become more dependent on innovation, for example, IoT and sensors, which will permit them to gather information continuously. It can influence the business execution of SCM and lay out the basis for the development and rise of AI in SCM. Extra possibilities for future examination might come from this review. For example, it thinks about the fundamental components for the use of AI in SCM. The examination of hierarchical and social components affecting the reception of an AI functional viewpoint in the SCM could be the subject of captivating future exploration. Regardless of the extraordinary potential that AI offers in SCM, it has the best approach before its actual worth is understood [14].

Market basket analysis is a bunch of computations envisioned to assist organizations with figuring out the fundamental buys in their stores. Certain reciprocal merchandise is frequently bought together, and the Apriori Algorithm can convert them into binary format.

Understanding how items to deal with can be utilized in everything from advancements to strategic pitching to suggestions. While the models given were fundamentally retail-determined, any industry can profit from a superior comprehension of how its items move [15].

Market basket methods couldn't care less what's in the crate or regardless of whether there is a container: they find affiliation designs on any things in a set. Things in a web-based truck and titles in a line are only instances of applying market crate examination. Merchandise in a market container, SKUs in the assembling and circulation parts of a production network, and even side effects, medications, and kinds of unfavorable medication occasions are issues that can be tended to with similar methods.

CONSENT FOR PUBLICATON

I, Gaddam Venkat Shobika, give my consent for the publication of identifiable details, which can include photograph(s) and/or videos and/or case history and/or details within the text ("Material") to be published in this chapter.

ACKNOWLEDGEMENTS

I thank my professor Dibya Nandan Mishra for their expertise and assistance throughout all aspects of our study and for their help in writing the chapter.

REFERENCES

[1]   Haris Gacanin, and Mark Wagner, "Artificial intelligence paradigm for customer experience management in next-generation networks: challenges and perspectives", *IEEE Network,* pp. 1-7, 2019.
[http://dx.doi.org/10.1109/MNET.2019.1800015]

[2]   S. Sha, and M. Nazim, "Creating a Brand Value and Consumer Satisfaction in E-Commerce Business Using Artificial Intelligence", *Proceedings of International Conference on Sustainable Computing in Science, Technology and Management (SUSCOM).,* 2019 https://ssrn.com/abstract=3351618
[http://dx.doi.org/10.2139/ssrn.3351618]

[3]   Q.N. Nguyen, and A. Sidorova, "Understanding user interactions with a chatbot: A self-determination theory approach", *Americas Conference on Information Systems,* 2018.

[4]   M.H. Huang, and R.T. Rust, "Technology-driven service strategy", *J. Acad. Mark. Sci.,* vol. 45, no. 6, pp. 906-924, 2017.
[http://dx.doi.org/10.1007/s11747-017-0545-6]

[5]   I. Tammela, A.G. Canen, and P. Helo, "Time-based competition and multiculturalism", *Manage. Decis.,* vol. 46, no. 3, pp. 349-364, 2008.
[http://dx.doi.org/10.1108/00251740810863834]

[6]   Y. Duan, J.S. Edwards, and Y.K. Dwivedi, "Artificial intelligence for decision making in the era of Big Data – evolution, challenges and research agenda", *Int. J. Inf. Manage.,* vol. 48, pp. 63-71, 2019.

[http://dx.doi.org/10.1016/j.ijinfomgt.2019.01.021]

[7]　M. Wollowski, R. Selkowitz, L. Brown, A. Goel, G. Luger, J. Marshall, A. Neel, T. Neller, and P. Norvig, "A survey of current practice and teaching of AI", *Proc. Conf. AAAI Artif. Intell.,* vol. 30, no. 1, 2016.
　　　[http://dx.doi.org/10.1609/aaai.v30i1.9857]

[8]　E. Brynjolfsson, and A.N.D.R.E.W. Mcafee, "Artificial intelligence, for real", *Harv. Bus. Rev.,* vol. 1, pp. 1-31, 2017.

[9]　K. Lamba, and S.P. Singh, "Big data in operations and supply chain management: current trends and future perspectives", *Prod. Plann. Contr.,* vol. 28, no. 11-12, pp. 877-890, 2017.
　　　[http://dx.doi.org/10.1080/09537287.2017.1336787]

[10]　A. McAfee, and E. Brynjolfsson, *Machine, platform, crowd: Harnessing our digital future.* WW Norton & Company, 2017.

[11]　M. Vahidi Roodpishi, and R. Aghajan Nashtaei, "Market basket analysis in insurance industry", *Management Science Letters,* vol. 5, no. 4, pp. 393-400, 2015.
　　　[http://dx.doi.org/10.5267/j.msl.2015.2.004]

[12]　R.V. Rodriguez, P.S. Sairam, K. Hemachandran, Ed., *Coded Leadership: Developing Scalable Management in an AI-induced Quantum World.* CRC Press, 2022.
　　　[http://dx.doi.org/10.1201/9781003244660]

[13]　H. Fatorachian, and H. Kazemi, "Impact of Industry 4.0 on supply chain performance", *Prod. Plann. Contr.,* vol. 32, no. 1, pp. 63-81, 2021.
　　　[http://dx.doi.org/10.1080/09537287.2020.1712487]

[14]　M. Chui, J. Manyika, M. Miremadi, N. Henke, R. Chung, P. Nel, and S. Malhotra, "Notes from the AI frontier: Insights from hundreds of use cases", *McKinsey Global Institute,* vol. 2, 2018.

[15]　"Machine learning for business analytics", In: *Real-Time Data Analysis for Decision-Making.* (1st ed.). Productivity Press: New York, 2022, p. 190.
　　　[http://dx.doi.org/10.4324/9781003206316]

CHAPTER 5

Data Insights by Using Data Visualization and Exploration

Choppala Swathi Priya[1,*], **Sai Santosh Potnuru**[1], **Ishank Jha**[1], **Hemachandran K.**[2] and **Chinna Swamy Dudekula**[3]

[1] *Woxsen School of Business, Woxsen University, Kamkole, Sadasivpet, Telangana-500078, India*

[2] *Department of Artificial Intelligence, School of Business, Woxsen University, Hyderabad, India*

[3] *Engineering and Environment (Advanced Computer Science with Advanced Practice), Northumbria University, Newcastle, United Kingdom*

Abstract: Any effort to make data more understandable by presenting it visually falls under the wide definition of data visualization. The graphic depiction of quantitative information is called data visualization. In other words, data visualizations turn big and small data sets into images that the brain can process more quickly. Users using data visualization can gain insight into vast volumes of data. They can use it to find new patterns and data mistakes. Users can concentrate on areas that show progress or warning signs by making sense of these patterns. This procedure then advances the business. Surprisingly frequently, data representations assume the well-known shape of charts and graphs in our daily lives. It can be used to uncover unknown facts and trends. Good data visualizations result when communication, data science, and design work together. When done properly, data visualizations provide important insights into huge, complex data sets in simple, understandable ways. Data visualization is the graphic depiction of information and data. Trends, outliers, and patterns in data are easy to spot and comprehend with the use of data visualization tools, which employ visual components like charts, graphs, and maps. Furthermore, it enables employees or business owners to convey information to non-technical audiences without misunderstanding them. In the world of big data, it is essential to have access to tools and technology for data visualization to analyze vast volumes of data and make data-driven decisions. We will discuss data visualization, its significance, data visualization technologies, and other topics in this article.

Keywords: Big Data, Data Visualization, Data Exploration, Outliers.

---

* **Corresponding author Choppala Swathi Priya:** Woxsen School of Business, Woxsen University, Kamkole, Sadasivpet, Telangana-500078, India; E-mail: swathipriya.choppala_2023@woxsen.edu.in

Hemachandran K., Raul V. Rodriguez, Umashankar Subramaniam & Valentina Emilia Balas (Eds.)

INTRODUCTION

The most crucial tool for data analysis is visualization. It offers the first line of defense by exposing complex data structures that cannot be comprehended in any other way. We can find unexpected consequences and work to contradict anticipated ones. Bits and bytes make up data, which is kept inaccessible in a file on a computer hard drive. We must first see the facts to comprehend them. I'll define "visualizing," which only refers to textual data representations, in this chapter. For instance, data visualization can be defined as the simple act of importing a dataset into spreadsheet software [1]. The unseen data suddenly appears on our screen as a visual "image," therefore, the inquiries should not be the question of whether analysts must visualize data or not, but rather whether the type of visualization could be most helpful in a particular circumstance. Or, to put it another way: when does going beyond table visualization make sense? The response is nearly always. Tables alone cannot provide us with a thorough picture of the dataset or help us to spot patterns in the data right away. Geographical patterns are the most prevalent example here, which are only visible after displaying data on a map other patterns, however, will be revealed in this chapter.

VISUALIZATION AS A TOOL FOR INSIGHT DISCOVERY

Expecting data visualization tools and methodologies to lower the threshold for pre-made tales from datasets is wildly unrealistic. There are no predetermined norms or guidelines that will ensure a tale. Instead, I think it makes more sense to search for "insights," which, in the hands of a talented author, may be deftly incorporated into the story. We'll learn something new about our data almost every time we create a new graphic. While some of the discoveries may be well-known (though possibly not yet confirmed), others may be brand-new or even startling to us. While some novel insights might signal the start of a narrative, others might simply be the consequence of data inaccuracies, which are most likely to be discovered by displaying the data.

I found the following procedure to be quite beneficial for improving the effectiveness of uncovering insights in data shown in Fig. (**1**):

How to Visualize Data?

The dataset is seen from a fresh viewpoint thanks to visualization. Data can be seen in a variety of ways [2].

When dealing with a limited amount of data points, tables are especially helpful. Together, they show their full potential by displaying labels, having the capacity to rank and sort data and amounts, and filtering data and amounts. Edward Tufte also recommended inserting minor chart elements, such as a single line chart or a bar chart per row, into table columns. Tables undoubtedly have their limitations, as was indicated in the opening. They are excellent for demonstrating outliers in one dimension.

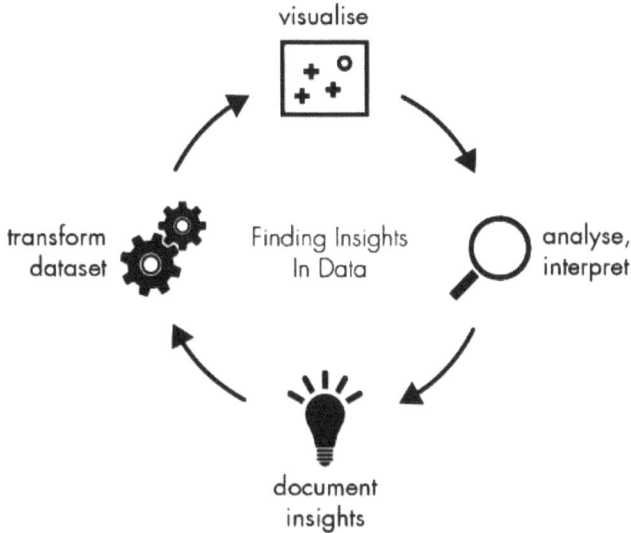

Fig. (1). Data Insights.

In Fig. (**2**), comparable to the top ten, they, however, perform poorly when evaluating numerous dimensions at once (For example, considering the population of each country over time).

In general, charts let you associate visual characteristics of geometric forms with dimensions in your data. The usefulness of specific visual characteristics has been extensively discussed; the gist is that color is challenging and position is crucial. For example, the x- and y-positions are mapped to two dimensions in a scatterplot. Even the size or hue of the displayed symbols can have a third dimension. Bar graphs are excellent for comparing categorical data, whereas Line charts are ideal for displaying temporal changes. Chart components can be stacked on top of one another. Using numerous iterations of the same chart to compare only a few groups in your data can be a very effective strategy (also known as small multiples). You can investigate different features of your data in any chart by using various scales (For example, linear or log scale).

In reality, the vast majority of the information we are working with has some connection to real persons. The ability of maps to connect data to the physical environment is what gives them power. Imagine a dataset of crimes that are geolocated. You must determine the locations of the crimes. Additionally, maps can show spatial relationships among the data, such as a north-south trend or a shift from urban to rural areas.

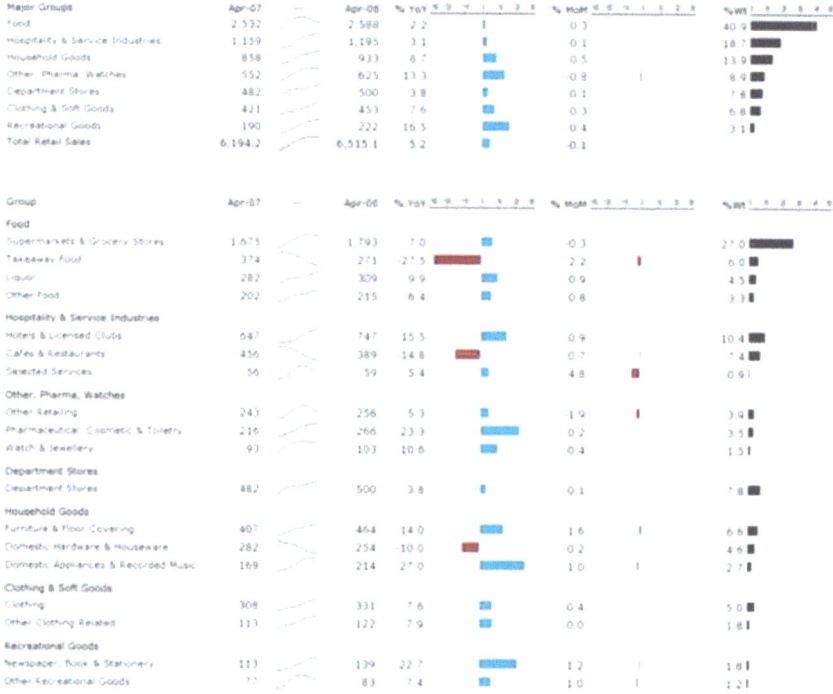

Fig. (2). Data Sorting [3].

The graph shown in Fig. (**3**) is the fourth most significant sort of visualization, speaking of relations. Using graphs, you may display how your data points are connected *via* their edges (nodes). The nodes' positions are subsequently determined by a certain amount of complicated graph design techniques, allowing us to notice the network's structure immediately. In general, the trick to graph visualization is to model the network suitably. Relationships are not always present in datasets, and even when they are, they might not be the most intriguing feature to examine. Determining the edges between nodes can be left to the journalist in some cases. The edges of the United States Senate Social Graph link senators who cast the same vote in over 65% of the votes cast, which serves as a superb illustration of this. Interpret what you see *via* analysis.

Following the visualization of your data, the following stage is to conclude the visuals you produced shown in Fig. (**4**). You might contemplate this:

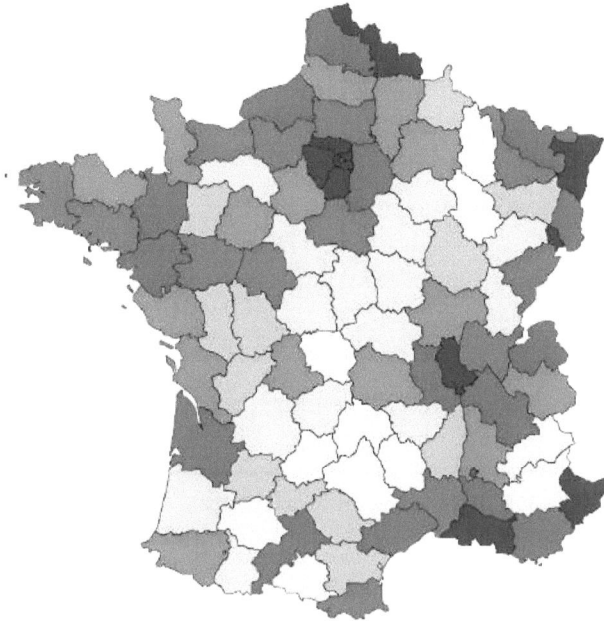

Fig. (3). Visualization of Data [3].

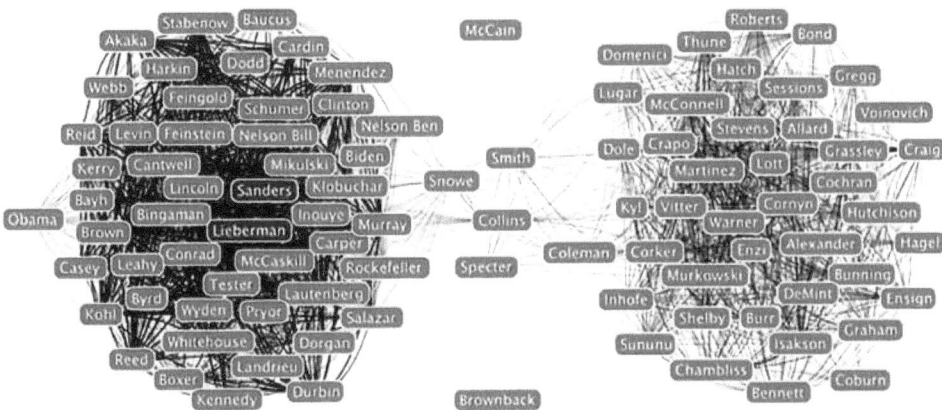

Fig. (4). Visualization through Graph [3].

- What can I make out in this picture? Is this what I anticipated?
- Are there any noteworthy trends?
- What does this signify in light of the available information?

Even while a visualization may be beautiful, there are occasions when it seems to reveal little about the data itself. But no matter how insignificant the visualization, there is nearly always something you can learn from it.

Transform the Data

Naturally, based on the knowledge you acquired from the last visualization, you can have an idea of what you want to see next. There is a chance that you noticed an intriguing trend in the dataset that you want to investigate further [4].

These modifications are possible:

Zooming: Focus on a specific graphical detail aggregation to consolidate numerous data elements into a single entity.

Filtering: Exclude data elements that are not relevant to our main focus (temporary).

Elimination of anomalies: To exclude single instances does not accurately 99% of the dataset represented. Consider that after visualizing a network, all that was left was a jumble of nodes connected by a large number of edges.

Consider the following: one frequent transformation step would be to filter some of the edges if you visualized a graph and received a tangle of nodes connected by hundreds of connections (a common outcome when visualizing so-called densely connected networks). If the edges, for instance, we may eliminate any financial transfers that are below a particular threshold from donor to receiving nations.

Tools

The problem of tools is not simple. Every data visualization tool on the market shines in a certain field. Data manipulation and visualization must be easy to use and reasonably priced. If altering the visualization options takes hours, you won't experiment as much. It doesn't follow that you shouldn't learn how to utilize the tool, though. However, it will be effective after you've learned it.

Choosing a technology that tackles both data manipulation and data visualization issues is frequently useful. Data import and export must be done frequently as a result of dividing the tasks among many instruments. The following is a brief list of various data manipulation and visualization tools:

Spreadsheet applications such as Google Docs, LibreOffice, and Excel.

Pandas and R (r-project.org) are statistical programming frameworks (pandas.pydata.org).

Quantum GIS, ArcGIS, and GRASS are examples of Geographic Information Systems (GIS).

Tools for Data Wrangling: Data Wrangler, Google Refine.

Visualization software that does not require programming, such as Many Eyes and Tableau Public.

Why a Tool Like Data Visualization is so Effective?

It makes sense intuitively to present a graph as a node-link structure, even to those who have never worked with graphs before.

Fast: This is possible only when data is presented concretely because our brains are excellent at seeing patterns. With visualization, we can quickly identify trends and outliers.

Flexible: Because the world is so interconnected, graph visualization is useful as long as your data contain any intriguing relationships.

Insightful: Compared to static display or raw data, interactive graph data exploration enables users to obtain more in-depth knowledge, comprehend the context, and ask more questions.

DATA VISUALIZATION AND BIG DATA

Big data [1] is a collection of organized, semi-structured, and unstructured information that businesses collect and use for advanced analytics initiatives like predictive modeling and machine learning. Big data processing and warehouse systems have developed into a typical element of data operation structures in businesses together with the technology that powers big data analytics. Doug Laney, an analyst with the consulting company Meta Group Inc. at the time, developed these qualities in 2001; Gartner made them well-known after acquiring Meta Group in 2005.

CHARACTERISTICS OF THE BIG DATA

Volume: The amount of data generated and retained. The volume of data also affects how valuable and illuminating it is and if it qualifies as big data. Large amounts of big data frequently exceed terabytes and petabytes [5, 6].

Variety: The change from structured to unstructured data presented difficulties for the technology and techniques already in use. The main objective of big data technology is to collect, store, and process significant amounts of data. Eventually, whether using big data or traditional RDBMSs, structured data processing remained optional.

Velocity: The speed at which data is generated and processed to meet the demands and challenges of development and progress. Big data access in real-time is typical. The production of big data exceeds that of little data. There are two different forms of velocity in big data: the frequency of generation and the frequency of handling, recording, and publishing.

Veracity: Data quality and data worth are related to how trustworthy or reliable the data is. For big data to be helpful for analysis, it must not only be abundant but also trustworthy. Recorded data may have varying levels of data quality, which makes accurate analysis challenging.

Value: The importance of the data that may be gleaned from processing and studying large datasets. A value could also be established by analyzing big data's numerous properties. The value could also represent how profitable the knowledge discovered through big data analysis is.

Variability: Big data's revolving door of sources, formats, and structures define it. Raw data from numerous sources are organized and unstructured data from various sources, and both can be combined in big data analysis. It is possible to turn unstructured data into structured data as part of the processing process.

Architecture

Big data architecture's [5, 7] objective is to manage data that is too complex or massive for traditional database systems. Different businesses might enter the big data realm at different points depending on the tool's capabilities. It would mean hundreds of gigabytes of data for some people and several terabytes for others.

The data environment has changed. Data has transformed what you can do with it and what you are expected to achieve. Some data come in quickly, necessitating continuous collection and monitoring. Other information comes in far larger amounts, but more slowly. How will you handle this data?

THE FOLLOWING WORKLOADS ARE AMONG THOSE THAT BIG DATA SOLUTIONS FREQUENTLY INCLUDE

• Batch processing of enormous data sources while idle.
• It is feasible to process huge data in motion in real-time.

- Interactive exploration of huge data.
- Machine learning and predictive analytics.

BIG DATA ARCHITECTURES SHOULD BE TAKEN INTO CONSIDERATION

Unstructured data must be changed before it can be analyzed and reported on since it is too large for a conventional database to store and process. Gather, process, and analyze unbounded data streams instantly or as soon as possible.

MAIN ADVANTAGES AND DISADVANTAGES OF BIG DATA

Advantages of Big Data

- Data quality has a direct influence on the effectiveness of business processes [5, 6].
- Mediocre quality seller data in the purchase-to-pay process might affect missing purchase contracts or price information, causing detainments in carrying critical particulars.
- Businesses are espousing big data results or algorithms to simply do what they have always done, performing with little data loss.

Disadvantages of Big Data

- A considerable portion of big data is unstructured, making it more expensive to store vast amounts of data using traditional storage.
- It has the potential to be exploited to manipulate customer records and may exacerbate social inequality.
- Rapid changes in large data might cause real-world values to differ.

BIG DATA AND DATA VISUALIZATION RELATIONSHIP

The demand for visualization has expanded as big data and data analytic activities gain popularity [1]. Businesses are utilizing machine learning more and more to collect vast amounts of data, which can be difficult to go through, analyze, and explain. This process may be sped up with the use of visualization, which would also help stakeholders and business owners understand the information being presented.

The ability to use facts to form opinions and images to tell stories is becoming more and more crucial for professionals. Due to the enormous amounts of data, big data visualization opens up new possibilities and challenges. To make the spectator more aware of the data volumes, various visualization approaches were created.

Pie charts, histograms, and business graphs are just a few examples of how big data visualization frequently goes beyond conventional visualization techniques. It substitutes more complex visualizations, such as heat maps and fever charts. Big data visualization needs the use of powerful computer systems to collect raw data, analyze it, and transform it into graphical representations that humans can readily utilize to derive conclusions.

WHILE BIG DATA VISUALIZATION HAS ITS BENEFITS, THERE ARE ALSO SOME SERIOUS DISADVANTAGES FOR ENTERPRISES. THE FOLLOWING ARE THEIR NAMES

- To make the most of huge data visualization technology, a visualization expert must be hired. To guarantee that businesses use their data to the fullest extent possible, this professional must be able to choose the relevant data sets and visualization strategies.
- IT and management are commonly involved in big data visualization initiatives due to the need for powerful computer technology, effective storage systems, and a move to the cloud.
- The accuracy of the insights that can be obtained *via* big data visualization depends on the data being displayed. As a result, it is crucial to have systems in place for managing and regulating the quality of corporate data, metadata, and data sources.

DATA EXPLORATION

Finding the most effective method to arrive at a result that a certain audience can understand is the process of data exploration [8]. Data exploration is crucial to the final visualization process since it considers the audience of the *viz* first and foremost by comprehending the various forms of perceptions and examining data based on these notions. For references, consider the following data exploration techniques: Businesses can explore data using a combination of automated and human techniques. Because these tools make it possible for users to rapidly and easily observe the majority of the pertinent elements of a data collection, analysts frequently employ automated tools like data visualization software for data exploration. Users can determine variables that are likely to produce amusing findings from this phase. Users can see if two or more variables correlate by presenting data graphically, for as through scatter plots, density plots, or bar charts, and decide if they are good candidates for additional study, which may include:

The analysis of just one variable is known as univariate analysis. The practice of analyzing two variables to ascertain their relationship is known as bivariate analysis. A type of study called multivariate analysis uses several different result variables. Analyzing principal components is the investigation of a larger set of potentially correlated factors and conversion of those variables into fewer, uncorrelated ones filtering and analyzing data in Excel spreadsheets or using programs to evaluate raw data sets are two manual data exploration techniques.

CONCLUSION

Although it is obvious that the topic of data visualization has many potential applications across a wide range of fields, we also need to be conscious of its practical and ethical challenges. Because the human brain is not designed to process such a large amount of unstructured, raw data and turn it into something usable and understandable, we require data visualization. We have outlined some crucial theoretical and practical guidelines for creating data visualizations. Additionally, we have explored and analyzed several data visualization instances. As we've seen, creating a data visualization that is both ethical and successful is a difficult process.

Data Visualization's Future

A new era in data visualization is beginning. The potential value that analytics and insights can offer is changing as a result of new intelligence sources, theoretical advancements, and improvements in multidimensional imaging, with visualization playing a critical role. The fundamentals of efficient data visualization remain constant.

The principles of good data visualization, such as substance, context, and acting ability, have received a lot of attention in the past. A quick review of the following principles seems appropriate as timeless principles that will always be crucial, regardless of medium or format:

Meaningful data visualization should be used. A lack of substance cannot be made up for by inventive pictures, which can boost retention and attention. "Every single pixel should testify directly to substance," said puritan Edward Tufte. The visualization needs to be accurate and pertinent. David McCandless' novel Billion Dollar O'Gram offers an example of how adding greater relevance can be accomplished by considering the bigger picture. Absolute numbers in a connected environment, according to McCandless, "don't give you the complete picture. They are not entirely accurate. For a more complete picture, we require relative figures that are linked to other pieces of information.

Above all, data visualization should facilitate decision-making, a difficult objective for many. While data and analytics are perceived to be more crucial for enterprises, producing meaningful insights continues to be a top difficulty, according to a new [KPMG study] (International 2015).

CONSENT FOR PUBLICATON

I, Choppala Swathi Priya, give my consent for the publication of identifiable details, which can include a photograph(s) and/or videos and/or case history and/or details within the text ("Material") to be published in this chapter.

ACKNOWLEDGEMENTS

I thank my professor, Dr. Hemachandran, for their expertise and assistance throughout all aspects of our study and for their help in writing the chapter.

REFERENCES

[1]   M. Islam, and S. Jin, "An Overview of Data Visualization", In: *2019 International Conference on Information Science and Communications Technologies (ICISCT)* IEEE: Tashkent, Uzbekistan, 2019, pp. 1-7.
[http://dx.doi.org/10.1109/ICISCT47635.2019.9012031]

[2]   "Machine learning for business analytics", In: *Real-Time Data Analysis for Decision-Making.* 1st ed.. Productivity Press, 2022, p. 190.

[3]   Available at: https://datajournalism.com/read/handbook/one/understanding-data/using-data-visualization-to-find-insights-in-data

[4]   R. Ramloll, C. Trepagnier, M. Sebrechts, and J. Beedasy, "Gaze data visualization tools: opportunities and challenges", In: *Proceedings. Eighth International Conference on Information Visualisation.* IEEE: London, UK, 2004, pp. 173-180.
[http://dx.doi.org/10.1109/IV.2004.1320141]

[5]   D. Keim, H. Qu, and K.L. Ma, "Big-Data Visualization", *IEEE Comput. Graph. Appl.,* vol. 33, no. 4, pp. 20-21, 2013.
[http://dx.doi.org/10.1109/MCG.2013.54] [PMID: 24921095]

[6]   P.K. Kotturu, and A. Kumar, "Data mining visualization with the impact of nature inspired algorithms in big data", In: *4th International Conference on Trends in Electronics and Informatics (ICOEI)(48184)* IEEE: Tirunelveli, India, 2020.
[http://dx.doi.org/10.1109/ICOEI48184.2020.9142979]

[7]   S.M. Ali, N. Gupta, G.K. Nayak, and R.K. Lenka, "Big data visualization: Tools and challenges", In: *2016 2nd International Conference on Contemporary Computing and Informatics (IC3I)* IEEE: Greater Noida, India, 2016, pp. 656-660.
[http://dx.doi.org/10.1109/IC3I.2016.7918044]

[8]   M.C.F. de Oliveira, and H. Levkowitz, "From visual data exploration to visual data mining: A survey", *IEEE Trans. Vis. Comput. Graph.,* vol. 9, no. 3, pp. 378-394, 2003.
[http://dx.doi.org/10.1109/TVCG.2003.1207445]

CHAPTER 6

Application of Computer Vision to Laboratory Experiments

P.K. Thiruvikraman[1,*], Devendra Dheeraj Gupta Sanagapalli[1] and Simran Sahni[1]

[1] *Department of Physics, BITS Pilani Hyderabad Campus, Telangana-500078, India*

Abstract: Computer vision has been applied in many fields. We demonstrate some simple applications of computer vision to improve the accuracy of laboratory experiments. The techniques used require only a camera in a mobile phone. Individual frames can be extracted from the video using PYTHON/MATLAB. Further processing of the images can be used to accurately measure the time period of oscillation or rotational time periods. The techniques described can be easily extended to a variety of fields.

Keywords: Computer vision, Colour, Digital Image Processing, MATLAB, Object Recognition, OpenCV.

INTRODUCTION: BACKGROUND

Undergraduate Engineering and Science students typically do a lot of experiments in the laboratory to hone their experimental skills. In most of these experiments, the measurements of quantities like length and time are done manually by the students. The measurement of time using a stopwatch is especially prone to errors as the experimenter has to react quickly to start or stop the clock. While manual measurements do help in building the experimental skills of the student, they are prone to errors. If the errors are significant, then the student's understanding of the subject will be affected negatively. It is of course possible to interface instruments to a computer using a GPIB interface for accurate measurement, but these are available only in research laboratories and not in undergraduate labs. The omnipresent mobile phone can be used to record videos of experiments, and it is possible to extract information from these videos. We describe two experiments, one involving a coupled pendulum, and the other involving a flywheel. While the measurement of a time period is crucial to both experiments,

---

Corresponding author P.K. Thiruvikraman: Department of Physics, BITS Pilani Hyderabad Campus, Telangana-500078, India; E-mail: thiru@hyderabad.bits-pilani.ac.in

Hemachandran K., Raul V. Rodriguez, Umashankar Subramaniam & Valentina Emilia Balas (Eds.)

the nature of the motion is different in the two is different. The coupled pendulum executes oscillatory motion, while the flywheel executes rotational motion. Image processing/computer vision techniques can then be used to deduce the period of oscillation or rotation, as the case may be. The accuracy of these computerised measurements is significantly better than manual measurements. The methods used here can be extended easily to analyse more complicated motions in real life situations.

Computer vision has been applied in many areas. We mention a few areas where it has been used extensively: biomedical image processing, remote sensing, character recognition, self-driving cars and unmanned vehicles.

Within biomedical imaging, it has been used for detection of various diseases like cancer [1], monkey pox [2], and COVID-19 [3]. Most of the current research in this area involves the use of artificial neural networks and deep learning. In work being reported here, we use very simple detection techniques, which mostly use colour for recognition.

THE COUPLED PENDULUM EXPERIMENT

The coupled pendulum consists of two pendulums which are usually thin rods coupled by a spring, as shown in Fig. (**1**).

To measure the time period of oscillations, we have to track the motion of the thin rods which are hanging vertically in Fig. (**1**). One way to track the motion of the rods is to use the fact that they are straight lines in the image. We can use line detection algorithms like the Hough transform [4] to locate the pendulums in the figure. However, the pendulums are supported by some rods, and the algorithm has to distinguish between the two rods. To avoid this complication, we pasted small coloured stickers on the pendulums (the stickers are coloured yellow in Fig. (**1**)).

The pendulums are set into oscillations, and a video of the experiment (oscillations) was recorded using the camera in a mobile phone. The individual frames in the video were extracted using the VideoCapture function, which is part of the OpenCV (PYTHON) library. Matlab function "VideoReader" [5] can also be used to extract the frames. Further processing of images was done using PYTHON. A white background was placed behind the pendulum (as shown in Fig. (**1**)) to avoid the presence of other objects in the background with a colour similar to the sticker. The video was also cropped to avoid other objects present in the background. The location of the coloured sticker was detected by convolving each frame with a mask of the required colour and size. The RGB values of the pixels were converted to HSV [6] as HSV is more sensitive to changes in light. Making use of the moments module of OpenCV, we marked the centroid of the sticker.

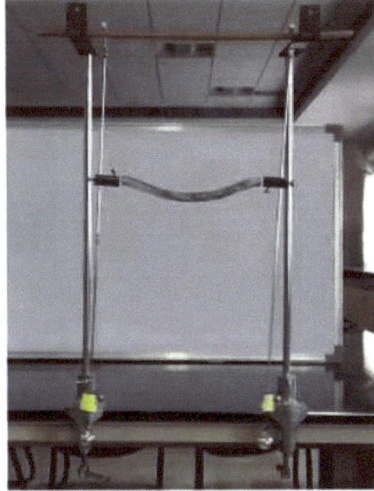

Fig. (1). The coupled pendulum.

The x coordinate of the centroid of the sticker is shown in Fig. (**2**) as a function of time.

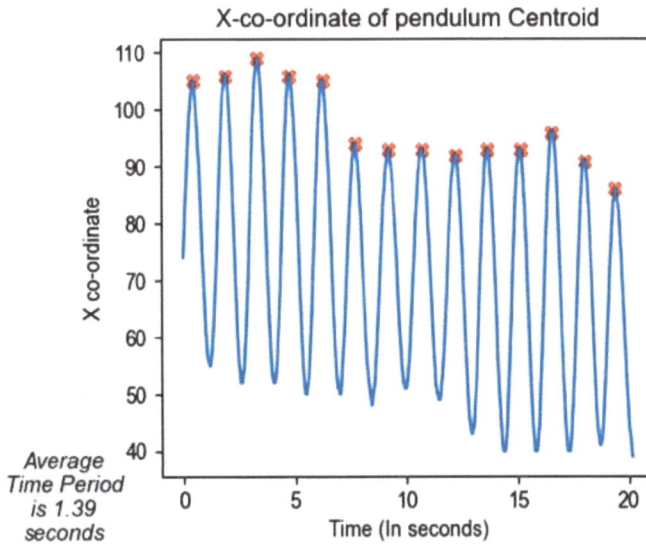

Fig. (2). The x coordinate of the centroid of the pendulum as a function of time.

Since a handheld camera was used, shaking of the hand can lead to a spurious change in the coordinates of the centroid (the sudden change in Fig. (**2**) after 10 seconds). However, this does not affect the determination of the time period.

The time period of the pendulum extracted from the data in Fig. (**2**) is given in Table **1**.

Table 1. Time period of the pendulum extracted from the data.

X coordinate in pixels	Time (seconds)	Time period (seconds)
106	1.91	1.47
109	3.35	1.44
106	4.76	1.41
105	6.23	1.47
94	7.67	1.44
93	9.18	1.51
92	10.69	1.51
92	12.2	1.51
92	13.64	1.44
93	15.11	1.47
96	16.52	1.41
91	17.99	1.47

In manual measurement, we usually record the time taken for 10 or 20 oscillations and calculate the average time period from the measurements. This is done to reduce the error due to the finite reaction time of the experimenter. This also means that any transient phenomena cannot be detected. However, Table **1** gives the time period for each individual oscillation.

The time interval between the 2 frames in the video is 0.03245 sec, which means the minimum time period/error in the time period is 0.0649 sec. Any variation beyond 0.0649 sec is an actual variation in the time period of the pendulum. Observed variation is within ±0.0649 sec, which rules out any variation in the time period due to transient external disturbances. The results obtained above show that the accuracy of these automated measurements is way beyond those obtained by manual measurements. The method adopted can be used to investigate oscillatory or other types of motion in many other systems.

The key steps used in our method are summarized in the flowchart given in Fig. (**3**).

```
┌──────────────┐      ┌──────────────┐      ┌──────────────┐
│ Video        │  →   │ Extraction   │  →   │ Detection of │
│ recording    │      │ of frames    │      │ all pixels of│
│              │      │ (MATLAB)     │      │ required     │
└──────────────┘      └──────────────┘      │ colour in    │
                                            │ each frame   │
                                            └──────────────┘
                                                    │
                                                    ↓
┌──────────────┐      ┌──────────────┐      ┌──────────────┐
│ Time difference│ ←  │ Identify     │  ←   │ Calculation  │
│ between       │     │ turning      │      │ of centroid  │
│ adjacent turning│   │ points    in │      │ of coloured  │
│ points gives time│  │ $x_{cm}$     │      │ paper        │
│ period        │     └──────────────┘      └──────────────┘
└──────────────┘
```

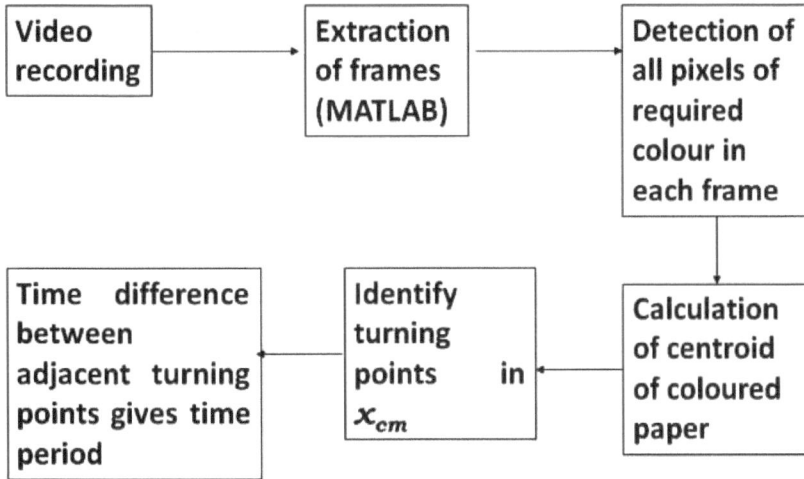

Fig. (3). Flowchart for the method.

THE FLYWHEEL EXPERIMENT

We next studied the flywheel experiment. The basic experimental set-up is shown in Fig. (**4**).

Fig. (4). The flywheel experiment.

The aim of this experiment is to determine the moment of inertia (also known as rotational inertia) of the flywheel by a measurement of the rotational period of the flywheel. The system consists of a disc mounted on a horizontal axle. A string is wound around the axle and a weight is hung from the end of the string. The weight exerts a torque on the flywheel, which gives an angular acceleration of the flywheel. The weight gets detached from the flywheel once the string has unwound completely. Applying conservation of energy, we get the following equation [7] for the moment of inertia of the flywheel:

$$I = \frac{Nm}{N+n}\left(\frac{2gh}{\omega^2} - r^2\right)$$

(1)

Here h is the height by which the weight falls before getting detached, N is the number of revolutions of the flywheel after the weight gets detached (it comes to a stop because of friction), n is the number of times the string is wound around the flywheel, r is the radius of the flywheel, and ω is its angular velocity at the point when the weight gets detached from it. Since it is practically difficult by manual means to directly measure the instantaneous angular velocity of the flywheel, we normally measure the average velocity (which should be equal to $\omega/2$) of the flywheel before it comes to a stop by noting down N and the time taken for it to come to a stop after the weight detaches. However, using a video of the flywheel experiment, it is, in fact, possible to measure the instantaneous angular velocity of the flywheel.

As in the coupled pendulum experiment, we used OpenCV in resizing the image frames, converting the images from RGB scale to Hue-Saturation-Value (HSV) scale, displaying image frames, detecting colored stickers on the flywheel and drawing bounding rectangles on image frames.

The time module in Python provides various time-related functions. We use this module to get the time at any desired instant. The module has a function time() that returns the number of seconds elapsed from January 1, 1970, 00:00:00 (UTC). We note down the time at the start of the video. Relative to this time, we can compute the time elapsed during the experiment at any desired instant of the recording.

Once we acquire the HSV values, we start reading through the video frame after frame. There is a possibility that we can encounter multiple colors in a single frame. Therefore, we have to identify all the required colors from the frame. The function decetColor() takes several arguments in which, one of them is the color that is to be detected in the image. The program then loops through the pixels of the image to find the pixels whose HSV values are in the specified range. If a particular pixel falls in this range, it is whitened and blackened if it's not.

Contours are found in this masked image which represents the area of our desired color in the frame. These contours are then highlighted on the image with a bounding rectangle. We can access the center, width, height and angle of rotation of this rectangle from inbuilt functions available in Python. Analyzing the center of this bounding rectangle gives us information whether the flywheel has completed a rotation or not.

Fig. (5) shows the bounding rectangle drawn around the coloured sticker, which helps us to locate the sticker.

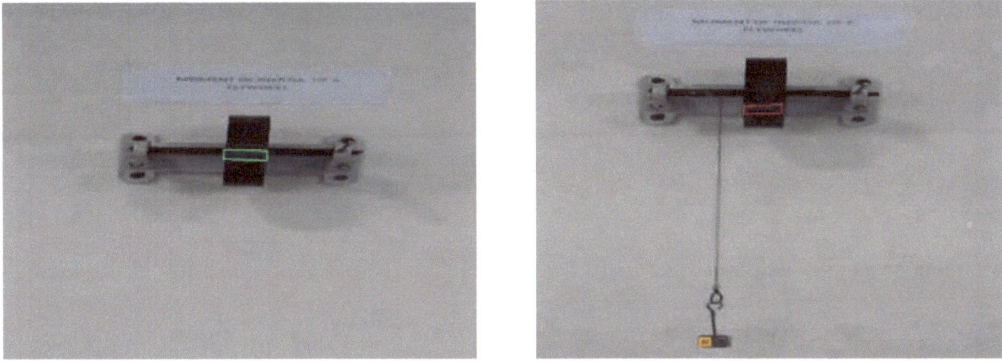

Fig. 5. Bounding rectangle around the coloured sticker.

We have selected a center point at the center of the flywheel. Whenever the green sticker passes through this center, we consider that a rotation has been completed. Unfortunately, the green sticker passes this center in between the frames of the video. To tackle this problem, we have defined a central region of width 10 pixels above and below this center point. Whenever the green sticker passes through this region, we note down the frame number of the video. Dividing the frame number with the fps of the video gives us the time elapsed from the start of the video.

To compute the time period of rotation, we subtract the time elapsed for the rotation and the time elapsed for the previous rotation. The time period of the ith rotation is governed by the formula:

$$T[i] = time[i] - time[i-1] \tag{2}$$

Where T[i] is the time period of ith rotation, time[i] is the time elapsed from the beginning since the start for i rotation to be complete. The angular velocity is computed from the time period from the following formula:

$$\omega = \frac{2\pi}{t} \tag{3}$$

The time at which the hanging weights touched the ground is manually chosen from the video. We then plot angular velocity as a function of time elapsed. We then return the maximum angular velocity recorded and the number of rotations of the flywheel to the driver code.

The angular velocity of the flywheel is plotted as a function of time in Fig. (6).

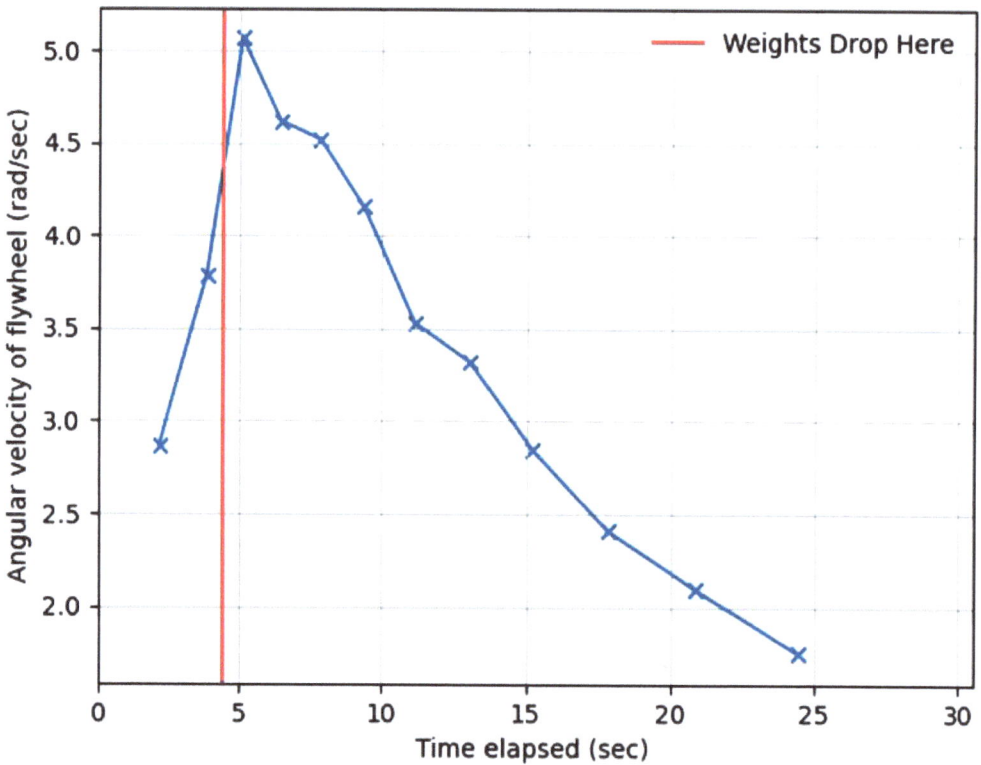

Fig. (6). Angular velocity of the flywheel a function of time.

From Fig. (6), we see that we can determine the angular velocity of the flywheel as a function of time, something which is practically impossible by manual means. We also note that the magnitude of the angular acceleration (slope of the graph) is higher before the weight drops when compared to the magnitude of the deceleration after the weight drops. We also see that angular velocity when the weight drops can be read off from Fig. (**5**) and substituted into equation 1 to determine the moment of inertia. Alternately the average angular velocity can be calculated from Fig. (**6**). We conclude that the automated system gives us more information when compared to the manual experiment and, therefore, will improve the conceptual understanding of the student.

CONCLUSION

Automated measurement of the time period of oscillation and rotation has been demonstrated in two standard laboratory experiments. The automation leads to greater accuracy and gives more information about the experiment than would be possible in manual measurement. The techniques adopted can be easily extended to any other experiment which involves tracking the motion of objects. In the present work, colour stickers have been used on the objects to facilitate the recognition process. Such an option is available in carefully controlled laboratory experiments but may not be available if the objects being imaged are not accessible (for instance, where images are being acquired by satellites). In such cases, shape and texture information may be more appropriate for detection of objects. While artificial neural networks exhibit superior performance in most situations, they also involve more computation. In the context of laboratory experiments, colour and shape information may be adequate. It is also possible to provide the features extracted (colour and shape) as inputs for artificial neural networks. Such an approach may lead to improved performance.

REFERENCES:

[1]   E. Jana, R. Subban, and S. Saraswathi, "Research on skin cancer cell detection using image processing", In: *International Conference on Computational Intelligence and Computing Research (ICCIC)*. IEEE: Coimbatore India, 2017, pp. 1-8.

[2]   Gulmez Burak, International Research in Engineering Sciences*A hybrid deep convolutional neural network model for monkeypox disease detection* Egitim Publishingpp. 49-64.

[3]   B. Gülmez, "A novel deep neural network model based Xception and genetic algorithm for detection of COVID-19 from X-ray images", *Ann. Oper. Res.*, pp. 1-25, 2022.
[http://dx.doi.org/10.1007/s10479-022-05151-y] [PMID: 36591406]

[4]   R.O. Duda, and P.E. Hart, "Use of the Hough transformation to detect lines and curves in pictures", *Commun. ACM*, vol. 15, no. 1, pp. 11-15, 1972.
[http://dx.doi.org/10.1145/361237.361242]

[5]   P.K. Thiruvikraman, *A Course on Digital Image Processing with MATLAB®*. Institute of Physics publishing, 2019.
[http://dx.doi.org/10.1088/978-0-7503-2604-9]

[6]   *Digital Image Processing, R.C.Gonzalez, R.E.Woods.* 2nd ed. Pearson, 2002.

[7]   Mishra Arun kumar, *Practical physics for engineers.* Laxmi publications, 2006.

CHAPTER 7

Violence Detection for Smart Cities using Computer Vision

Jyoti Madake[1,*], Shripad Bhatlawande[1], Abhishek Rajput[1], Aditya Rasal[1], Sambodhi Umare[1], Varun Shelke[1] and **Swati Shilaskar[1]**

[1] *Vishwakarma Institute of Technology, Pune, Maharashtra, India*

Abstract: There is a need for developing deep learning solutions to analyze videos to identify any violence being present. This paper proposes a method for the detection of the presence of violent activities in videos using Deep Neural Networks. Recently there has been a rapid development happening in the field of Deep Neural networks, but the number of solutions that have been developed for violence detection is very few. The proposed solution will play a major role in transforming the way law enforcement works and support the government's initiative to make cities smarter. The model is built using CNN for video frame feature extraction and LSTM to capture localized features present in the video frames. The LSTM extracts the localized features using the spatiotemporal relationship between the video frames. The local motion present in the video is analyzed. This work focuses on accuracy and fast response time. The performance was evaluated on the hockey fight dataset to detect violent activities.

Keywords: CNN, LSTM, Violence detection.

INTRODUCTION

Public violence is a major threat to a country's economic and social well-being. There has been a significant rise in cases of public violence, theft, *etc.,* in India. Sometimes peaceful protests can turn into a more aggressive and hostile nature leading to fights between groups, protestors causing vandalism, or even attacking the police authorities. This can lead to the destruction of public and private property and, in worst cases, loss of human life too. India is experiencing massive economic growth and is becoming an emerging economic giant; Events like these are a major threat to the sustainability of this economic growth. Public violence also hampers India's image in the world. India experiences many instances of public violence and theft which can only be solved with innovation in policing methods. Large-scale and small-scale losses of public and private property can be

* **Corresponding author Jyoti Madake:** Vishwakarma Institute of Technology, Pune, Maharashtra, India;
E-mail: Jyoti.madake@vit.edu

avoided if these events are detected early on and stopped when instances of minor violence are detected. India is turning towards making its cities smarter, leading to efficiency in all areas of governance. In the smart cities initiative, a mere installation of surveillance systems does not solve our issue. Because the manpower in the police force is limited and therefore a lot of incidents may go unnoticed in real-time, leading to delayed justice and giving time to conspirators to flee. This issue of limited manpower leads to significant errors in detecting violence in real-time and can be solved by automating the process. Computer vision-based solutions have become popular for building images and videos to solve such image and video data-based problems. Some of the computer applications in various fields, including:

Healthcare: Computer vision is used in medical imaging to help doctors diagnose diseases and detect anomalies [16]. It can also be used to track the progression of diseases and to monitor patients remotely.

Automotive industry: Computer vision is used in driver assistance systems, such as collision warning and lane departure warning systems. It is also used in self-driving cars to help them navigate roads and avoid obstacles.

Retail industry: Computer vision is used in retail to analyze customer behavior and to personalize shopping experiences. It is also used in inventory management to track products and prevent theft.

Agriculture: Computer vision is used in agriculture to monitor crop growth, detect diseases and pests, and optimize irrigation.

Security and surveillance: Computer vision is used in security and surveillance to detect suspicious behavior, recognize faces, and track objects.

Entertainment: Computer vision is used in the entertainment industry to create special effects, enhance visual effects in movies and games, and to track the movement of actors in motion capture.

Robotics: Computer vision is used in robotics to help robots navigate and interact with their environment, recognize objects, and perform tasks autonomously. Overall, computer vision has a wide range of applications in various areas, and its use is growing rapidly as technology advances.

In this article, we have projected a computer vision-based solution to detect the presence of violence. The violence detection method is backed by deep learning. Violence detection requires features that have space and time domain correlation between them. The solution to detect violence is intended to help local police stations. Therefore, the proposed system needs to have a low computing power, and the solution needs to be fast, accurate, robust, and require low computing power. The proposed system is built using a combination of (CNN) as our feature extractor and the LSTM for spatiotemporal feature extraction from video frames. This model was tested on the hockey fight dataset as of now, having 1000 videos, out of which 500 had instances of violence and 500 were non-violent.

LITERATURE SURVEY

Most earlier efforts [1 - 4] used a technique to extract features that had time and space domain correlation. These researchers have used feature Motion Scale-Invariant Feature Transform (MoSIFT) [1] and Space-Time Interest Points (STIP) [2]. The STIP is the most prevalent action descriptor. MoSIFT discovers distinctive local features using local appearance and movement. MoSIFT detects the local feature points and then encodes these interest points to include the motion present, The reference [1] mentions the comparison between STIP and MoSIFT feature description techniques. It was observed that MoSIFT has better accuracy of 91% on the hockey dataset. Another reference [3] compares STIP and SIFT. It mentions that STIP has superior performance compared to SIFT.

In most cases, a deep learning approach is employed to extract Spatio-temporal information on its own. T. Hassner *et al.* proposed a deep learning architecture to extract the spatiotemporal information from the videos. The Deep neural network structure is structured in a sequence of convolutional layers, followed by multilateral integration activities to identify discriminatory key points. The convLSTM - Convolutional Long-Short Memory Encoder is used to find the frame changes to identify the violent scene in the videos, encoding frame level changes that define violent scenes in videos. This model trained and tested on the hockey dataset achieved 97.1% accuracy, which is the greatest accuracy. This model is the fastest model [4].

The reference paper [4] proposes a technique using irregular motion information. The proposed method calculates the object's velocity vector in the image. The 8-dimensional quantization of the estimated motion vector. The Co-occurrence in the direction of the quantized motion vector is then calculated. The proposed technique efficiently detects violent events, according to their experimental data. The reference paper [5] proposes a violence detector that consists of four types. The first module divides a video into several scenes. The second is tasked with extracting the skin and blood-colored regions from the scene frames, which are then further processed and filtered to find the regions of interest (which may be anything) that equate to violent stuff . After that, the motion intensities of these potential locations are calculated, and those with high values are assigned to the violent class.

It's not easy to implement their strategy for real-time processing. On the other hand, Hendel *et al.* [6] present an effective and probabilistic method. However, they had assumed that the video scene c ould be described with numerous space-time tubes, each of which contains a moving object in the scene. In crowd videos, this condition is frequently impractical.

As the outreach of computer vision and artificial intelligence is increasing, many new domains are finding their implementation in this technology. But every new application has to overcome some hurdles due to data insufficiency in a specific context. Transfer learning allows the use of training data from different domains. In such cases, if the knowledge transfer is carried out successfully, the learning performance can be greatly improved without the need for expensive data labeling efforts [7]. Speed of detection can be improved by recognition of kinematics in the video, which is obvious during a violent interaction between individuals. Random transform to the consecutive frames is applied to estimate the random accelerators [8]. Action MACH Filter was suggested for action detection, unlike feature tracking methods used in much other literature [9]. It has demonstrated that, under some conditions, for reliable poses, less than ten video frames are necessary. Categorization of actions such as sub-second delays is taken into account [10]. CNN and LSTM are the networks that internally use the Convolutional neural network [11] is the best suitable local feature extractor. The most efficient extractor of all kinds of handmade techniques for extracting local features and then extracting features feeds in LSTM Layer to learn temporary relationships rather than using any separation layer like ANN or any other method of learning and differentiation.

In-depth reading was also used in the study of the video feature in the unattended area [12]. In the study [12], the authors use the stacked ISA to study the Spatio-temporal features of videos. Although this method has shown good results in action recognition, it is still very computer-assisted in training and difficult to develop for testing on large databases. Deep neural networks are very difficult to train. The study [13] introduced a residual learning framework to facilitate the training of more in-depth networks than previously used. The study [14] presented a transfer learning based efficient model for the classification of images. It is built on bidirectional convolutional LSTM, which uses spatiotemporal encoder. This encoder offers dual-guided coding and high elementwise integration of the code [15].

The literature review shows the systems implemented for scene recognition. The systems studies were data intensive and computationally complex. This paper uses transfer learning to reduce dependency on datasets. It uses LSTM to identify the scene.

METHODOLOGY

This section provides an overview of the methodology executed while implementing the project. This paper presents a scene detection methodology since it predicts violence in the video. Fig. (**1**) shows a screenshot of a violent scene in the video dataset, and Fig. (**2**) shows the non-violent scene. The paper attempts to solve the problem of scene/event classification and detection to detect and convey violent scenes. The architecture of the project involves using CNN layers and LSTM. The layers of CNN extract the feature information from the input frames and the sequence prediction of these feature frames is done using the LSTM layer. The workflow of the solution in a simplified manner is video input, feature extraction, scene classification, detection, and prediction, as represented in the flowchart in Fig. (**3**). The methodology includes two major paths, CNN based feature extraction and violent action binary classification using LSTM.

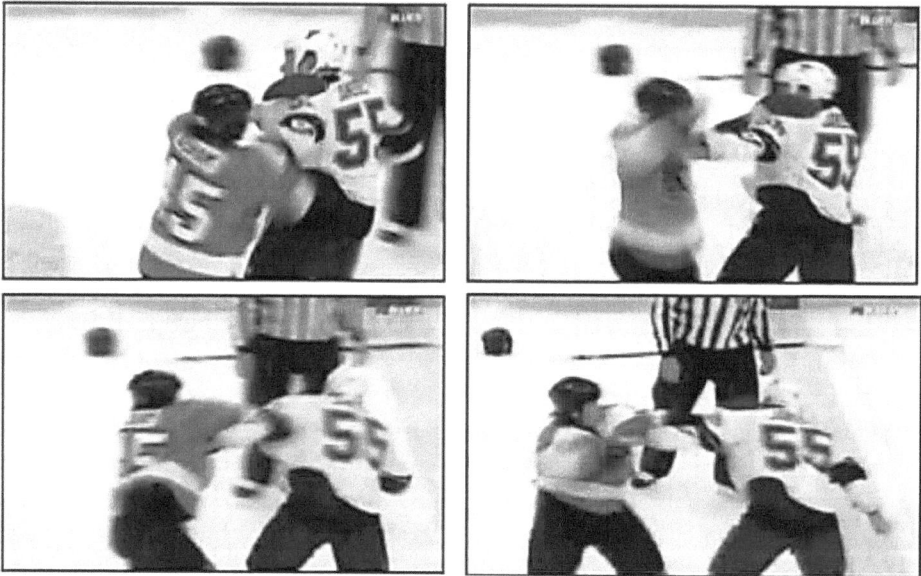

Fig. (1). Violent scenes in the dataset.

Fig. (2). Non - Violent scene in the dataset.

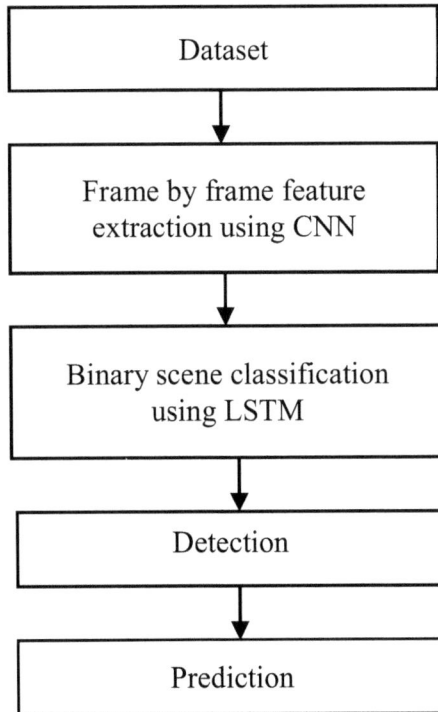

Fig. (3). System flowchart.

In feature extraction, a video is inputted and split into images/frames. The authors used a hockey fight dataset, as shown in Fig. (1) which contained 500 instances of violent scenes and 500 instances of non-violent scenes. The dataset contained videos 2 seconds long, and 20 frames were obtained with fps of 10. These frames were then passed into a pre-trained CNN network with backbone model VGG16 in order to extract spatial features and obtain the transfer values.

Then the transfer knowledge is used to construct and train an LSTM decoder model, which learns in the temporal domain, interprets the features across time steps, and proper training, validation, as well as testing of the dataset, are also done, and finally, the work is put to test by actually predicting if the contents of video have any violence or not. The system flowchart is presented in Fig. (3). The various processing operations taking place on the data set have been described in the following section of the paper.

Feature Extraction

This process reduces the computing power required to process large datasets having a large number of variables by obtaining the best features from the dataset. These features require low computational power to be obtained, yet they describe the data with accuracy while maintaining originality. The process significantly helps in extracting useful information in the project as there are 20 frames per video to be analyzed, with the total number of videos equaling 1000. The paper uses CNN for feature extraction from video frames. CNN detects important features using multiple convolutional layers. CNN reduces the number of parameters of the feedforward type neural network without losing any useful feature information from the images. It does not require any monitoring with good computing power. It uses special convolution and integration functions and performs parameter sharing.

VGG is an advanced convolutional neural network model developed in 2014 by researchers at Oxford's Visual Geometry Group, giving the model its name VGG. It is one of the best-performing CNN models and is known for its simplicity. In this work, VGG16 is used, which is a 16 layers neural network with 13 layers of convolution and 3 layers of fully connected with 5 layers of max-pooling. The layer structure is shown in Table 1.

It consists of stacked convolution, which uses only 3*3 convolutions, with pooling layers followed by fully connected ANN. The authors used a pre-trained VGG16 model, trained over 15 million high-resolution images known as the ImageNet dataset. Three dense layers or fully connected (FC) layers are used for classifying videos. The results were saved in the form of transfer values just before entering the final layer of VGG16. Cache-file was used to store these resultant output values because when all the dataset videos were processed by passing through the VGG16 convolution layers, the transfer values were stored and passed to the LSTM network. The input of the VGG16 net had the dimensions: (224, 224). The output of the selected layer of VGG16 net had dimensions of 4096 per frame.

Transfer Learning

Transfer learning is a faster technique with a focus to store the learning parameters and knowledge obtained for one task set and using that knowledge to solve another related task. This method saves time and significantly improves performance and speed. The CNN model extracts features of one image at a time and transforms input pixels to the internal matrix or vector representation. Since, the project is concerned with scene detection and classification, which falls under the spatio-temporal domain, the problem cannot be solved by solutions at a spatial level. Therefore, the authors came up with a solution by building an LSTM model for learning features in the temporal domain and then predicting the sequence. The LSTM builds the internal state and updates weights by receiving the output from the CNN model in the form of a vector representation. At this step, transfer learning comes into the picture to transfer the learned feature values to the LSTM model. In the presented work, the transfer values were stored in RAM to save time and then transferred to the LSTM model.

Binary Classification using LSTM

The recurrent Unit (RU) is the basic building block of RNN. Out of multiple variants of recurrent units, such as the rather clunky LSTM and the somewhat simpler GRU which are used in this model. Experiments in the literature suggest that the LSTM and GRU have roughly similar performances. Even simpler variants also exist and the literature suggests that they may perform even better than both GRU and LSTM, but these variants are not implemented in Keras.

Whenever a new input is received, the internal state of the recurrent neuron gets updated automatically, which acts as a memory element. However, it is not a traditional kind of computer memory that stores bits that are either on or off. Instead, the recurrent unit stores floating-point values in its memory state, which are read and written using matrix operations so the operations are all differentiable. This means the memory-state can store arbitrary floating-point values (although typically limited between -1.0 and 1.0), and the network can be trained like a normal neural network using Gradient Descent.

Table 1. VGG16 Layer summary.

Layer (Type)	Output Shape	Parameter
Input	224, 224, 3	0
Conv blk 1-1	224, 224, 64	1792
Conv blk 1-2	224, 224, 64	36928
2D Max Pool-1	112, 112, 64	0
Conv blk 2-1	112, 112, 128	73856
Conv blk 2-2	112, 112, 128	147584
2D Max Pool-2	56, 56, 128	0
Conv blk 3-1	56, 56, 256	295168
Conv blk 3-2	56, 56, 256	590080
Conv blk 3-3	56, 56, 256	590080
2D Max Pool-3	28, 28, 256	0
Conv blk 4-1	28, 28, 512	1180160
Conv blk 4-2	28, 28, 512	2359808
Conv blk 4-3	28, 28, 512	2359808
2D Max Pool -4	14, 14, 512	0
Conv blk 5-1	14, 14, 512	2359808
Conv blk 5-2	14, 14, 512	2359808
Conv blk 5-3	14, 14, 512	2359808
2D Max Pool-5	7, 7, 512	0
Flatten – 1D	25088	0
Fully Connected 1	4096	102764544
Fully Connected 2	4096	16781312
Classification predict	1000	4097000
Total parameters: 138,357,544		

MODEL TRAINING AND OPTIMIZATION

The dimensions of the transfer learned values from CNN are used to define the LSTM architecture. A generalized LSTM unit has a cell, input, output, and forget gates. It analyses the data coming inside the network using only the relevant information and discards the unnecessary information. VGG16 converts each frame into a vector of 4096x1. From each video, we are processing 20 frames, so we will have 20 x 4096 values per video. The 20 frames of each video are considered for classification. The dimension of the LSTM input is 4096x20.

VGG16 results in two models traning.h5 and testing.h5 from the training dataset and testing dataset, respectively. Out of 800 videos from our training model, 750 videos remain in our training set, and the remaining 50 videos go into the validation set.

The main significance of the validation set is to authenticate the performance of the model during the training. This process offers data that allows us to analyze the hyperparameters and configurations of the model. The problem of overfitting is also prevented. LSTM is then trained using these two data. The total number of epochs used was 200. Three activation functions, relu, softmax, and sigmoid were used for maintaining non-linearity in the model.

In the backpropagation, to reduce the loss that occurred between mapping input features to the output classification, a loss function of mean squared is used. The testing has been done with 20% of the total videos. These videos have not been used to train the network. The model handles a random video for prediction. It looks for any violent frame and classifies the videos based on the two classes available in the dataset, violent and non-violent. The model proposed in this paper has accomplished an accuracy of 94% on the Hockey Fight Dataset.

RESULTS AND DISCUSSION

Fig. (4) shows the graph of accuracy in training databases and verification during training sessions. Fig. (5) shows the model loss in training and validation data sets during training sessions. Epoch uses collections, and each sample of the training database is transmitted back and forth *via* the neural network.

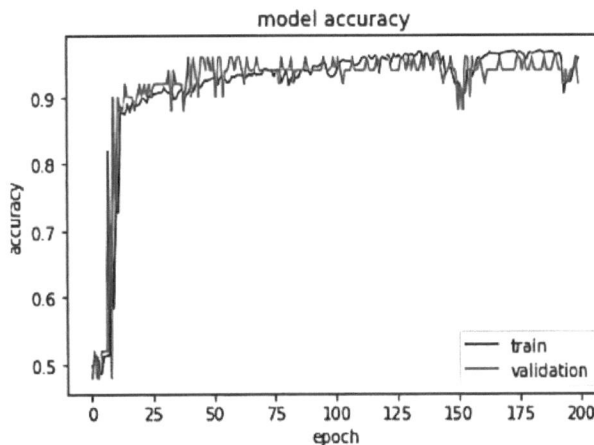

Fig. (4). Model accuracy during optimization.

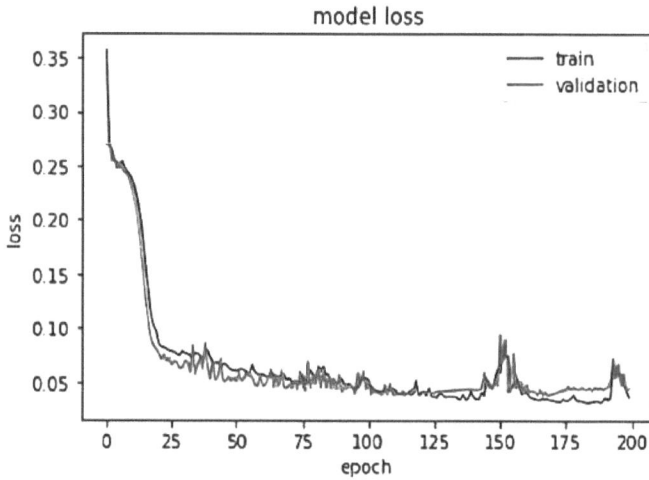

Fig. (5). Model loss during optimization.

Epoch count is also important because a lesser number of epochs may result in under fitting. After all the neural network has not learned enough. The training database needs to be passed multiple times or is required multiple times.

Fig. (**5**) shows the model loss during the optimization process. Fig. (**6**) shows the loss factor during optimization. For the third epoch, the model weights saved resulted in an accuracy of 98% which outperformed the highest known accuracy of 94%. The speed of the model also outperforms the speed of the fastest model for violent activity detection, which is around 131 frames per second.

Fig. (6). Violence detection and textual notification.

If any violent frame is detected, it prints, "Violence detected in the video", else, "No violence detected in the video" as shown in Fig. (**6** & **7**), respectively.

Fig. (7). No violence condition detection and textual notification.

If more data could be gathered for such classes, the model could be trained to perform better in both violent detection accuracy as well as crowd violence detection situations.

CONCLUSION

This paper presents a real-time violence detection model using CNN and LSTM. The proposed model was successfully trained and implemented with 94% accuracy, which was found to be at par or better than the previous works done in this area. This project has made a major advance in the fight against crime by making the detection of violence automated, real-time, fast, and accurate. A case should be made to incorporate this work into the law enforcement of our country. Further, more studies should be undertaken to increase the accuracy by building more advanced models. The accuracy rate of the violence detection system can be improved by using the various movies dataset, which will consist of more than a thousand violent and non-violent video short clips so that it can detect violent scenes with the same accuracy between two people as well as crowd violence detection.

REFERENCES

[1]   E.B. Nievas, O.D. Suarez, G.B. Garc'ıa, and R. Sukthankar, "Violence detection in video using computer vision techniques", In: *International Conference on Computer Analysis of Images and Patterns* Springer, 2011, pp. 332-339.
[http://dx.doi.org/10.1007/978-3-642-23678-5_39]

[2]   S. Sudhakaran, and O. Lanz, "Learning to detect violent videos using convolutional long short-term

memory", In: *14th IEEE International Conference on Advanced Video and Signal Based Surveillance (AVSS)* IEEE, 2017, pp. 1-6.Lecce, Italy
[http://dx.doi.org/10.1109/AVSS.2017.8078468]

[3]   T. Hassner, Y. Itcher, and O. Kliper-Gross, "Violent flows: Real-time detection of violent crowd behavior", In: *IEEE Computer Society Conference on Computer Vision and Pattern Recognition Workshops..* IEEE, Providence, RI, USA, 2012.
[http://dx.doi.org/10.1109/CVPRW.2012.6239348]

[4]   F.D. De Souza, G.C. Chavez, E.A. do Valle Jr, and A.A. Araujo, "Violence detection in video using Spatio-temporal features", In: *23rd SIBGRAPI Conference on Graphics, Patterns and Images..* IEEE, Gramado, Brazil, 2011.
[http://dx.doi.org/10.1109/SIBGRAPI.2010.38]

[5]   S. Hochreiter, and J. Schmidhuber, "Long short-term memory", *Neural Comput.,* vol. 9, no. 8, pp. 1735-1780, 1997.
[http://dx.doi.org/10.1162/neco.1997.9.8.1735] [PMID: 9377276]

[6]   K. Simonyan, and A. Zisserman, "Very deep convolutional networks for large-scale image recognition", *arXiv,* 2014.

[http://dx.doi.org/10.48550/arXiv.1409.1556]

[7]   S.J. Pan, and Q. Yang, "A Survey on Transfer Learning", *IEEE Trans. Knowl. Data Eng.,* vol. 22, no. 10, pp. 1345-1359, 2010.
[http://dx.doi.org/10.1109/TKDE.2009.191]

[8]   "Fast violence detection in video", In: *Proceedings of the 9th International Conference on Computer Vision Theory and Applications..* IEEE, Lisbon, Portugal, 2014.

[9]   M.D. Rodriguez, J. Ahmed, and M. Shah, "Action mach: A Spatio-temporal maximum average correlation height filter for action recognition", In: *IEEE Conference on Computer Vision and Pattern Recognition.* IEEE, Anchorage, AK, USA, 2008.
[http://dx.doi.org/10.1109/CVPR.2008.4587727]

[10]   A.M.R. Abdali, and R.F. Al-Tuma, "Robust Real-Time Violence Detection in Video Using CNN And LSTM", In: *2nd Scientific Conference of Computer Sciences (SCCS).* IEEE, Baghdad, Iraq, 2019.
[http://dx.doi.org/10.1109/SCCS.2019.8852616]

[11]   O. Deniz, I. Serrano, G. Bueno, and T. Kim, "Fast violence detection in video", In: *2014 International Conference on Computer Vision Theory and Applications (VISAPP)..* IEEE, pp. 478-485, Lisbon, Portugal, 2014.

[12]   M. Ramzan, A. Abid, H.U. Khan, S.M. Awan, A. Ismail, M. Ahmed, M. Ilyas, and A. Mahmood, "A Review on State-of-the-Art Violence Detection Techniques", *IEEE Access,* vol. 7, pp. 107560-107575, 2019.
[http://dx.doi.org/10.1109/ACCESS.2019.2932114]

[13]   K. He, X. Zhang, S. Ren, and J. Sun, "Deep residual learning for image recognition", *Proceedings of the IEEE Conference on ComputerVision and Pattern Recognition,* IEEE, pp. 770-778, 2016.
[http://dx.doi.org/10.1109/CVPR.2016.90]

[14]   M. Hussain, J.J. Bird, and D.R. Faria, "A study on cnn transfer learning for image classification", In: *UK Workshop on Computational Intelligence* Springer: Cham, 2018, pp. 191-202.

[15]   A. Hanson, K. Pnvr, S. Krishnagopal, and L. Davis, "Bidirectional Convolutional LSTM for the detection of violence in videos", In: *Proceedings of the European Conference on Computer Vision (ECCV) Workshops.* Springer, 2018, pp. 280-295.

[16]   Burak Gulmez, "MonkeypoxHybridNet: A hybrid deep convolutional neural network model for monkeypox disease detection", In: *International Research in Engineering Sciences.,* Kamanli Mehmet, Ed., Egitim Publishing: Konya, 2022, pp. 49-64.

CHAPTER 8

A Big Data Analytics Architecture Framework for Oilseeds and Textile Industry Production and International Trade for Sub-Saharan Africa (SSA)

Gabriel Kabanda[1,*]

[1] *School of Business, Woxsen University, Hyderabad, India*

Abstract: Among the most revolutionary technologies are Big Data Analytics, Artificial Intelligence (AI) and robotics, Machine Learning (ML), cybersecurity, blockchain technology, and cloud computing. The research was focused on how to create a Big Data Analytics Architecture Framework to increase production capability and global trade for Sub-Saharan Africa's oilseeds and textile industries (SSA). Legumes, shea butter, groundnuts, and soybeans are significant crops in Sub-Saharan Africa (SSA) because they offer a range of advantages in terms of the economy, society, and the environment. The infrastructure, e-commerce, and disruptive technologies in the oilseeds and textile industries, as well as global e-commerce, all demand large investments. The pragmatic worldview served as the foundation for the Mixed Methods Research technique. This study employed a review of the literature, document analysis, and focus groups. For the oilseeds and textile sectors in SSA, a Big Data analytics architectural framework was created. It supports E-commerce and is based on the Hadoop platform, which offers the analytical tools and computing power needed to handle such massive data volumes. The low rate of return on investments made in breeding, seed production, processing, and marketing limits the competitiveness of the oil crop or legume seed markets.

Keywords: AI, Big Data Analytics, Cybersecurity, E-Commerce, Hadoop, Machine Learning, Oilseeds, Textile industry.

INTRODUCTION

Massive amounts of data are produced in the Internet of Things (IoT) age from a number of heterogeneous sources, such as mobile devices, sensors, and social media. Among the most revolutionary technologies are Big Data Analytics, Artificial Intelligence (AI) and robotics, Machine Learning (ML), cybersecurity, blockchain technology, and cloud computing. The automatic analysis of massive

* **Corresponding author Gabriel Kabanda:** School of Business, Woxsen University, Hyderabad, India;
E-mails: Gabriel.Kabanda@woxsen.edu.in / gabrielkabanda@gmail.com

Hemachandran K., Raul V. Rodriguez, Umashankar Subramaniam & Valentina Emilia Balas (Eds.)

data sets and the development of models for the broad relationships between data are the two key components of machine learning (ML). Data analytics [1] is "the act of analyzing huge amounts of data from many sources and in numerous variants to gain insight that can enable real-time or near-real-time decision making". The study [2] defines Big Data Analytics as the methods and technologies that can be used to examine vast quantities of complex data in order to enhance a company's performance. Technology-based analysis of large data is used to show social growth trends because big data enables precise analysis, which facilitates informed decision-making and productive work. (Fig. **1**) illustrates the five qualities of big data: volume, value, diversity, velocity, and truthfulness.

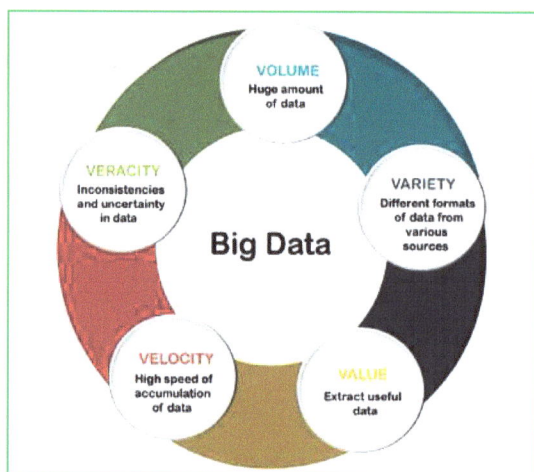

Fig. (1). Big Data Characteristics.

Legumes, shea butter, groundnuts, and soybeans are significant crops in Sub-Saharan Africa (SSA) because they offer a range of advantages in terms of the economy, society, and the environment. Sub-Saharan Africa produces a relatively little part of the world's agricultural output while having over 13% of the world's population and about 20% of its land area being used for agriculture [4].

Background

Sub-Saharan Africa (SSA) is home to more than 950 million people, or about 13% of the world's population. By 2025, oilseed production in SSA is projected to rise by 2.3 percent annually to 11 Mt, or just 2% of global production.

Southern Africa accounts for the biggest share of increased protein meal use in absolute volumes, even if the predicted growth is less dramatic at 16 percent due to a larger base. The fastest rates of growth are anticipated in Southern (1.4 percent annually) and Eastern Africa (1.2 percent annually) through 2025. As cattle industries grow in the coming years, protein meal consumption is rising throughout the majority of SSA, with Western Africa (43 percent) and Eastern Africa (43 percent) witnessing the highest increases (32 percent). By 2025, oilseed production in SSA is projected to rise by 2.3 percent annually to 11 Mt, or just 2% of global production. However, it is anticipated that overall imports into SSA will increase at a pace of 3.7 percent annually, with the majority coming from Nigeria (4 percent), Sudan (5 percent), Ethiopia (6 percent), and Kenya (3 percent). One of the commodities in the area with the quickest growth over the past ten years has been per capita consumption, which has increased at a pace of 2.1 percent annually. Sub-Saharan Africa's net food imports are predicted to increase over the next ten years, but investments that raise productivity may be able to buck this trend. Even though agricultural productivity has greatly grown over the past ten years, SSA continues to be the region with the greatest food insecurity and the slowest progress toward ending hunger. Table **1** below shows the global oilseed supply and distribution in million metric tons from 2017 to 2022.

Table 1. Major Oilseeds World Supply and Distribution (2017-2022) [million metric tons].

-	2017/18	2018/19	2019/20	2020/21	2021/22
Production	-	-	-	-	-
Copra	5.78	5.82	5.7	5.59	5.86
Cottonseed	45.25	42.97	43.55	40.81	42.75
Palm Kernel	18.69	19.46	19.32	19.03	20.05
Peanut	47.15	46.71	48.14	50.25	50.29
Rapeseed	75.28	72.85	69.6	73.59	71.18
Soybean	343.74	362.44	340.15	368.12	349.37
Sunflowerseed	48.01	50.66	54.2	49.25	57.38
TOTAL	583.9	600.91	580.65	606.64	596.87
Imports	-	-	-	-	-
Copra	0.13	0.2	0.15	0.08	0.08
Cottonseed	0.87	0.73	0.81	0.83	0.97
Palm Kernel	0.18	0.16	0.14	0.15	0.17
Peanut	3.08	3.53	4.34	4.31	4
Rapeseed	15.72	14.64	15.71	16.66	13.97
Soybean	154.11	146.02	165.12	165.47	154.46
Sunflowerseed	2.38	2.89	3.34	2.73	2.2

(Table 1) cont.....

-	2017/18	2018/19	2019/20	2020/21	2021/22
TOTAL	176.47	168.17	189.61	190.24	175.86
Exports	-	-	-	-	-
Copra	0.16	0.18	0.28	0.1	0.13
Cottonseed	0.89	0.84	0.88	0.96	1.16
Palm Kernel	0.16	0.07	0.08	0.06	0.05
Peanut	3.51	3.83	4.95	4.89	4.64
Rapeseed	16.53	14.62	15.92	17.98	13.84
Soybean	153.27	148.97	165.21	164.51	155.57
Sunflowerseed	2.75	3.24	3.66	2.91	2.59
TOTAL	177.28	171.75	190.97	191.41	177.97
Crush	-	-	-	-	-
Copra	5.67	5.83	5.56	5.52	5.71
Cottonseed	33.73	32.75	33.62	31.95	33.2
Palm Kernel	18.62	19.42	19.29	19.01	20.08
Peanut	18.15	18.05	19.24	19.86	20.1
Rapeseed	68.45	68.03	68.41	71.45	70.2
Soybean	295.44	298.61	312.31	315.08	313.68
Sunflowerseed	44.17	46.52	49.31	45.13	47.34
TOTAL	484.24	489.2	507.73	508	510.31
Ending Stocks	-	-	-	-	-
Copra	0.12	0.1	0.05	0.05	0.07
Cottonseed	1.96	1.82	1.61	1.41	1.42
Palm Kernel	0.23	0.26	0.24	0.25	0.23
Peanut	5.16	5.08	4.67	4.89	4.33
Rapeseed	8.14	9.93	7.81	5.96	4.27
Soybean	99.84	114.19	94.66	99.91	85.24
Sunflowerseed	2.79	2.57	2.92	2.56	7.61
TOTAL	118.24	133.95	111.96	115.02	103.16

Source: https://apps.fas.usda.gov/psdonline/circulars/oilseeds.pdf.

From Table **1**, the world production of oilseeds for the period 2017-2022 is shown in Fig. (**2**).

World Production of Oilseeds (2017-2022)

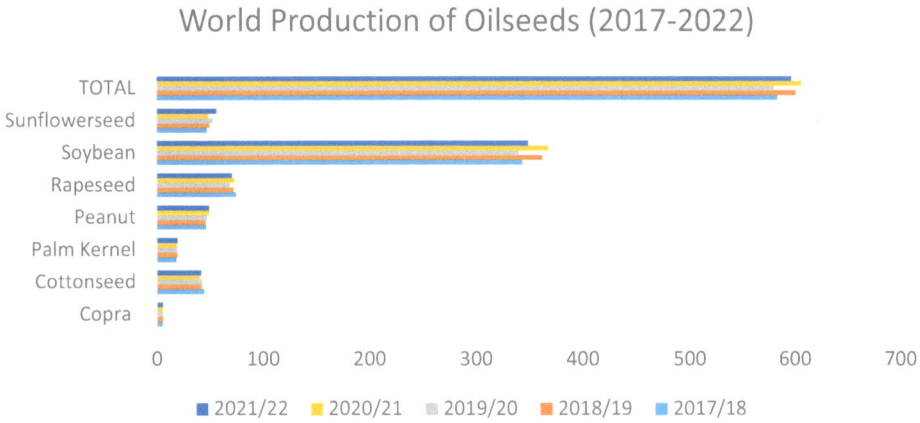

Fig. (2). World Oilseeds Crust Distribution (2017-2022).

The world oilseeds crush distribution for the period 2017-2022 is shown in Fig. (3).

World Oilseeds Crush (2017-2022)

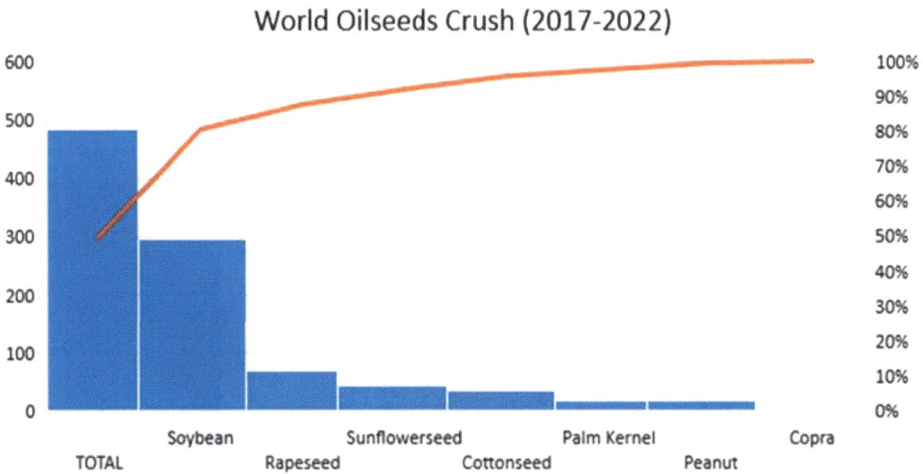

Fig. (3). World Oilseeds Crust Distribution (2017-2022).

The researchers' main concern was how to exploit and apply Big Data to enhance Sub-Saharan African output of textiles, oilseeds, and other agricultural products (SSA).

The top 15 textile exporters in Sub-Saharan Africa (SSA) are shown in Table **2** and illustrated in Fig. (**4**).

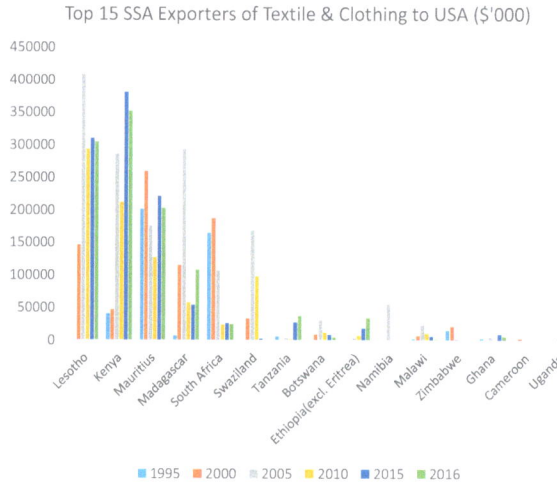

Fig. (4). The Top 15 SSA Exporters of Textiles and Clothing to US (US$'000).

Table 2. Top 15 SSA Exporters of Textiles and Clothing to US (US$'000). Source: World Bank.

	1995	2000	2005	2010	2015	2016
Lesotho	-	146365.92	408337.98	293625.99	310412.35	304867.13
Kenya	40557.59	46921.64	286480.04	212267.49	381118.11	352218.08
Mauritius	201844	259609	175787.13	127105.49	221933.63	203340.45
Madagascar	7475.2	115429.39	293757.75	58139.23	54429.66	108345.99
South Africa	164868.09	187000.1	107985.72	23786.08	26942.7	25108.16
Swaziland	-	33407.42	168769.77	97887.4	2807.2	1067.87
Tanzania	6084.74	253.87	4437.83	2159.59	27999.56	37883.39
Botswana	-	9028.59	31459.14	12209.52	8685.86	4981.05
Ethiopia(excl. Eritrea)	971.4	30.98	3829.68	7113.17	18799.72	34457.11
Namibia	-	196.09	56050.93	47.06	230	122.43
Malawi	2509.89	7653.83	24018.24	10728.07	6437.02	1603.53
Zimbabwe	15484.16	21574.02	3086.21	87.37	130.48	99.08
Ghana	3216.37	718.84	5749.01	1071.03	9620.28	6631.52
Cameroon	-	2769.28	407.24	749.97	1003.44	342.41
Uganda	-	5.07	5143.94	461.64	73.47	78.62

It is possible to raise the competitiveness of many textile and apparel inputs now produced in SSA countries by new or increased investment or other means, especially as the output of these inputs is frequently constrained and declining. The industry may be able to sustain or increase current levels of production and export of these inputs as well as increase the potential for new product creation with the aid of new or increased investment as well as other activities.

This paper aims to develop a Big Data Architecture framework for oilseeds and textile industry production and international trade for SSA.

STATEMENT OF THE PROBLEM

The African Growth and Opportunity Act (AGOA), a non-reciprocal trade preference program, was established by the US Congress in 2000 to assist developing SSA nations in improving their economies through increased exports to the US. Notably, the "third-country fabric clause" in AGOA permits US clothing imports from specific SSA nations to qualify for duty-free treatment even if the clothing products use yarns and fabrics manufactured by non-AGOA members, such as China, South Korea, and Taiwan. Furthermore, AGOA trade preferences offer much bigger duty savings for man-made-fiber products, which are subject to higher U.S. tariffs, even though SSA nations generate largely cotton-based textile and garment inputs due to a plentiful availability of local cotton. The most competitively produced cotton products in SSA nations seem to be cotton yarn, cotton knit fabric, denim fabric, and to a lesser extent cotton woven shirting fabric, either for direct export to or use in subsequent apparel production for export to the United States, the EU, and similar markets. However, the underdevelopment of the manmade-fiber textile and apparel sectors in most SSA countries prevents this. Numerous SSA industry sources claimed that the region produced textile and garment inputs for both local consumption and export beyond the region, suggesting that some local and regional markets may be competitive for all of these products.

Research Aim

The goal of this study is to develop a Big Data Architecture framework for the production of oilseeds, textiles, and international trade for SSA.

Research Objectives

The following are some of the research objectives:

1. To determine the competitive challenges facing Sub-Saharan Africa (SSA) in the production of oilseeds and the textile industry.

2. To evaluate yield production capacity and competitive variables across SSA for both oilseeds and textile industry.
3. To develop a Big Data Analytics architecture framework for usage by organizations in the oilseeds and textile production industry, as well as international trade in SSA.
4. To identify areas of Big Data applications that can help the oilseeds and textile industries in SSA increase their production and worldwide commerce.

Research Questions

1. What are the competitive issues in the oilseed and textile industries in Sub-Saharan Africa?
2. What has been the state of production capacity and competitive factors in the oilseeds and textile industries across SSA?
3. How do you create a Big Data Analytics architecture framework to assist and improve oilseed, textile, and international trade production?
4. What are some Big Data applications that can help SSA boost its oilseed and textile output and international trade?

Literature Survey

Generally the term "big data" refers to the rapidly expanding volume and velocity of data sets that are being accessible and connected. According to studies, big data may generally be defined using the four (4) V's of big data. The five properties of volume, value, diversity, velocity, and veracity are frequently used to describe big data, which is a collection of data from various sources. Big data analytics, which some academics define as the capacity to compile and analyze those fine-grained data sets, is already altering how insurers see sizable client bases, manage risks, and meet the diverse needs of their clients. The study [3] defines big data analytics as the straightforward application of analytics approaches to significant data sets. The five properties of volume, value, diversity, velocity, and veracity are frequently used to describe big data, which is a collection of data from various sources. Many significant businesses employ software for machine learning, artificial intelligence, data mining, cybersecurity, and other big data. Big data analytics, which some academics define as the capacity to compile and analyze those fine-grained data sets, is already altering how insurers see sizable client bases, manage risks, and meet the diverse needs of their clients. Literature [3] defines big data analytics as the straightforward application of analytics approaches to significant data sets.

Following a brief explanation, the sorts of analytics applicable to the oilseeds and textile industries are depicted in Fig. (**5**).

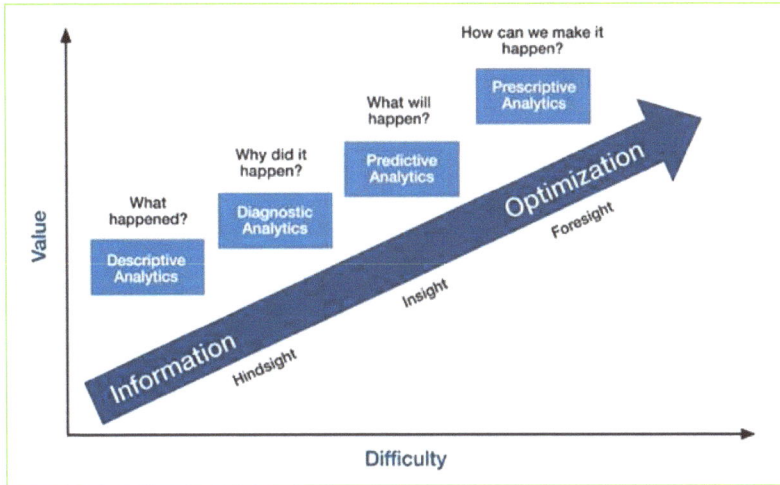

Fig. (5). Types of Analytics.

a. *Descriptive analytics:* Descriptive analytics is the process of describing and analyzing historical data that has been gathered about students, teachers, researchers, policies, and other administrative procedures. To report on current trends, patterns from samples are to be found.

b. *Predictive analytics:* Predictive analytics can give businesses data-driven insights and smarter decisions. By analyzing trends, finding links between linked problems, and detecting potential hazards or opportunities in the future, predictive analytics seeks to estimate the possibility of future events. Data linkages that would not be obvious using descriptive models, such as demographics, could be revealed with predictive analytics.

c. *Prescriptive analytics:* Based on reliable and consistent forecasts, prescriptive analytics enables enterprises to evaluate their current condition and choose the best possible course of action. It integrates analytical results from both descriptive and predictive models to consider evaluating and figuring out new ways to function in order to accomplish desired results while balancing restrictions. It showed that prescriptive analytics gives decision makers the ability to glimpse into the future of their crucial processes, identify possibilities, and provide the best course of action to quickly take advantage of that foresight.

Machine Learning (ML) is a method for instructing computers to learn. Big data analysis is known to be automated using machine learning, which also creates models of the fundamental relationships in the data. The way we teach, learn, and study in the educational setting could be completely changed by ML. Localization, transcription, text-to-speech, and personalisation are just a few of the ways that machine learning is expanding the reach and impact of online learning content [5]. Data mining can be handled through machine learning [6]. states that there are three different kinds of ML:

1. *Supervised learning*, in which training examples are provided to the techniques as inputs labeled with associated outputs;
2. *Unsupervised learning*, in which the methods are provided with unlabeled inputs;
3. *Reinforcement learning:* This type of learning employs data in the form of action sequences, observations, and rewards.

Machine learning is a big data analytics technique that mainly entails programming the creation of analytical models [7].

Finding anomalies, trends, and correlations in massive data sets in order to anticipate events is known as data mining [8]. The process of employing computers and automation to search through large data sets for patterns and trends, then converting those discoveries into business insights and predictions, is known as data mining. Data mining is a crucial component of data analytics and one of the core areas of data science, where advanced analytics methods are applied to draw out valuable information from massive data sets. Both are useful for identifying patterns in huge data sets, but they operate in quite different ways. Finding patterns in data is a technique known as data mining. Additionally, even though data mining is occasionally used in the machine learning process, it does not require constant human involvement *(e.g.,* a self-driving car relies on data mining to determine where to stop, accelerate, and turn).

Artificial Intelligence (AI) refers to computer systems that replicate human cognitive functions, including learning, reasoning, and self-correction. The aspect of AI that most closely resembles the thinking process of the human brain is its capacity to make decisions based on information rather than a preconceived set of instructions. In order to be considered artificial intelligence (AI), a system must be able to mimic human behavior and cognitive processes, as well as to capture and retain human knowledge and expertise.

Because of this, cyber security has emerged as a crucial idea in daily life, and understanding it is essential to averting cyberattacks on individuals and systems. The internet has opened up new prospects for all nations worldwide with the emergence of a global and borderless information culture, since technology plays a crucial role in social and economic development [9]. The internet has opened up new prospects for all nations worldwide with the emergence of a global and borderless information culture, since technology plays a crucial role in social and economic development [9]. According to a study [10], cyber security refers to techniques used to protect computer systems, networks, and software applications from cyberattacks. The major goal of the cyber security concept is to safeguard data integrity and confidentiality while also ensuring data accessibility when required. However, public concern about cyber security issues like social engineering and phishing varies as the nature of cyber threats does.

Infrastructure is the cornerstone of Big Data architecture. Having the appropriate tools for data storage, processing, and analysis is essential for every Big Data effort [11].

OILSEEDS AND TEXTILE PRODUCTION COMPETITIVE CHALLENGES IN SSA

The SSA countries were all faced with the following competitive issues, which are mentioned below.

• A Lack of Demand from the Apparel Industry

The market demand for textile and apparel inputs must be steady in order to support capital expenditures that take longer to pay off than apparel investments. A robust and successful garment sector provides this need.

• Lack of Understanding of Local and Global Market Opportunities

Many business professionals cited a dearth of marketing and networking opportunities, both domestically and abroad. Industry sources claim that the USAID has contributed to the growth of regional and global market potential, but further help is required.

• Insufficient Availability of Dependable Electricity at Affordable Prices

The cost of production is increased by the fact that several countries in the SSA region have some of the highest electricity bills in the world as well as unpredictable electrical supplies. Outages of electricity also result in worse quality and less efficient production of yarn and fabric.

• A Lack of Infrastructure for the Treatment of Waste Water and Clean Water

Clean water is a necessity for the production of yarn and fabric, especially for finishing and dyeing processes, yet many nations lack access to it. Inadequate transportation networks inside SSA further impede intraregional trade.

• A Lack of Competitive Access to Capital

When funding is readily available, the high cost of capital deters new investment in yarn, fabric, and other inputs and also drives up the price of current manufacturing. The finished goods produced on this equipment, especially the woven textiles, are frequently of insufficient quality to be exported to the United States, the EU, or other similar markets, or to be used in the fabrication of commercial clothing for export to these markets.

• Lack of Professional or Trained Labor

Industry sources claim that there is a lack of qualified labor in the textile and apparel sectors, particularly in countries without a sizable manufacturing base.

CONCEPTUAL FRAMEWORK FOR ADOPTION OF BIG DATA ANALYTICS

The study is guided by the conceptual framework shown in Fig. (**6**):

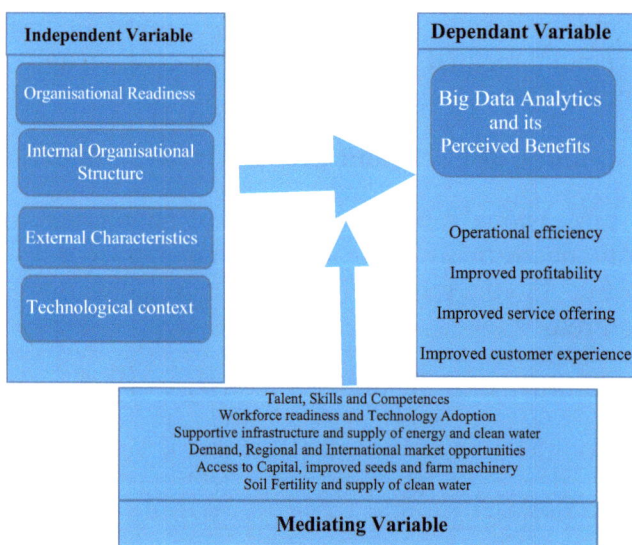

Fig. (6). Conceptual Framework.

Methodology

The Research Methodology explains the research philosophy of the study as well as the research design, research approach, data collection tools, target population, sampling method, and data processing techniques. The layers are represented by the research philosophy, approach, strategy, choice, time horizon, and techniques and processes in Fig. (**7**).

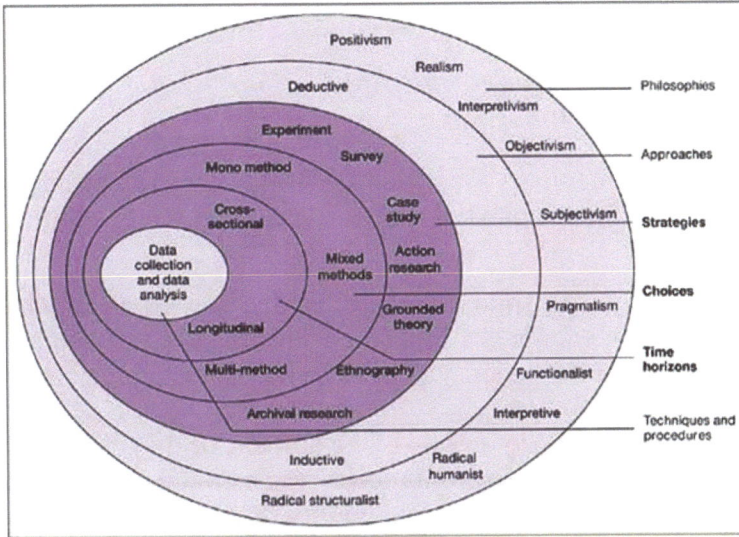

Fig. (7). Research Onion ([12]:138).

The Mixed Method Research (MMR) and the Pragmatism paradigm utilized in this study are closely related on a philosophical level. A worldview or paradigm known as pragmatism ought to guide the majority of mixed-methods studies. It is a problem-focused attitude that holds that the best research techniques contribute most significantly to the solution of the research topic. When conducting social science research, this frequently entails combining quantitative and qualitative methodologies to assess various facets of a research subject. The pragmatic worldview served as the foundation for the Mixed Methods Research technique. A mixed-methods strategy was used in this study, combining qualitative (Focus Group discussions) and quantitative techniques (a questionnaire). System logs, document analysis, and a literature review were also utilised in this study.

The goal of the focus group discussion was to conduct research, identify, and implement strategies for using big data to enhance Sub-Saharan Africa's output of textiles, oilseeds, and other crops, as well as global trade (SSA). The University of Zimbabwe's Masters class of 2021 assigned 10 groups of students to conduct surveys and interview the management of various corporations operating in the oilseeds and textile industries in Zimbabwe and other nearby Southern African nations. These groups served as the foundation for the focus groups. For analysis, secondary data were gathered from the US Department of Agriculture (https://apps.fas.usda.gov/psdonline/circulars/oilseeds.pdf), the World Bank, and the FAO (FAOSTAT, www.faostat.org).

Results and Discussion

Critical Challenges around the Applications of Big Data Analytics in the Oilseeds and Textile Industries

The focus group members identified the following critical obstacles around the deployment of big data analytics in the oilseeds and textile industries:

a. *Lack of Talent* - Despite the high demand for experienced data specialists, there is just not enough supply. Unfortunately, because there are still few data science degrees, the majority of elite colleges have not yet attempted to fill this gap.
b. *Storage and Scalability Issues* - The amount of data being produced exceeds the capabilities of the Big Data technologies that are now available. This can result in serious problems and compel systems to slow down or crash, which negatively affects the user experience and lowers the quality of the analysis.
c. *Security* - Security methods need to be updated to take into consideration the volume of data that Big Data employs in its analysis because they were not designed for a world with large amounts of data. The majority of businesses in

the oilseeds and textile industries are still in the early phases of developing their cybersecurity systems.

IMPLEMENTATION OF AN AI CHATBOT AND E-COMMERCE

All corporations and entities engaged in the oilseeds and textile industries must make significant investments in the infrastructure, which includes AI chatbots and e-commerce. A crucial component of AI chatbots is machine learning, which enables them to learn from their experiences and get better over time. Cybercash, Electronic Data Interchange (IDE), electronic advertising, business to company and business to customer online transactions can all be implemented by an organization using electronic business on a global scale. On a worldwide scale, small enterprises can compete with well-established and capital-rich companies thanks to internet commerce, wise planning, and effective policy-making. When it comes to business operations, electronic business refers to the use of technology to automate workflow and commercial transactions. Online delivery of digital content, electronic fund transfers, electronic share trading, electronic bills of lading, commercial auctions, collaborative design and engineering, online sourcing, public procurement, direct consumer marketing, and after-sales services are just a few examples of transactions in the global information economy.

The following are eight essential distinctive characteristics of required e-commerce technology:

1. Usability
2. Global impact
3. Universality
4. In-depth Information
5. Interactivity
6. Information content
7. Modification/personalization
8. Technology social

Every corporation operating in the oilseeds and textile sectors is urged to make significant investments in e-commerce and AI chatbots in order to provide the groundwork for the effective implementation of a big data analytics framework.

DATA ANALYSIS OF OILSEEDS PRODUCTION IN SSA

Soybean

For at least one million smallholder farmers in Africa, soybeans are an essential crop. The need for domestic processing to fulfill expanding domestic demand for soybean meal, particularly for the poultry feed industry, and the promising prognosis for edible oil are other reasons that have contributed to the increase in soybean demand. In the upcoming ten years, the top three soybean exporters—the United States, Brazil, and Argentina—will continue to account for around 90% of all exported soybeans, soybean meal, and soybean oil.

Fig. (8) below displays the global, Sub-Saharan African (SSA), and other countries' soybean output levels and yields per hectare. The stated increase in soybean output in has been driven roughly equally by area growth and yield growth. With yearly growth rates of 3% in area and 3.5 percent in yield, both area expansion and yield growth have contributed roughly equal amounts to the reported growth in soybean output in SSA.

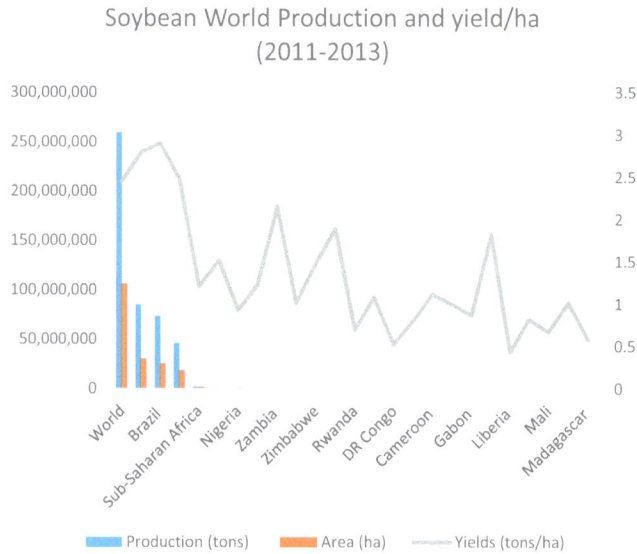

Fig. (8). Soybean world production and yield/ha.

Groundnuts

Groundnut ranks as the fifth-most significant oilseed crop in the world, behind oil palm, soybean, rapeseed, and sunflower. A significant oil, food, and feed legume crop is grown on 25.44 million hectares, yielding 45.22 million tons of pods in 2013; groundnut is grown in over 100 countries. Despite Africa's shrinking peanut market share, the crop nevertheless contributes significantly to export earnings in a number of nations (8% in Senegal and over 84 percent in Gambia in 2002, for example). Nutritious foods like groundnuts can boost rural residents' health. Groundnut haulms are used for hay and silage production, as well as fresh and dry cow feed. They include 8–15% protein, 1-3% lipids, 9–17% minerals, and 38–45% carbohydrates.

Cowpea

Sub-Saharan Africa (SSA) accounts for about 95% of the world's cowpea production, with West Africa contributing more than 80% of the continent's total. In Nigeria, cowpea is predominantly grown by low-income households, who stand to gain from cowpea research and extension because they generate more than 65 percent of the country's cowpea. Fig. (**9**) of the fundamental study depicts the yield production per ha and cowpea production.

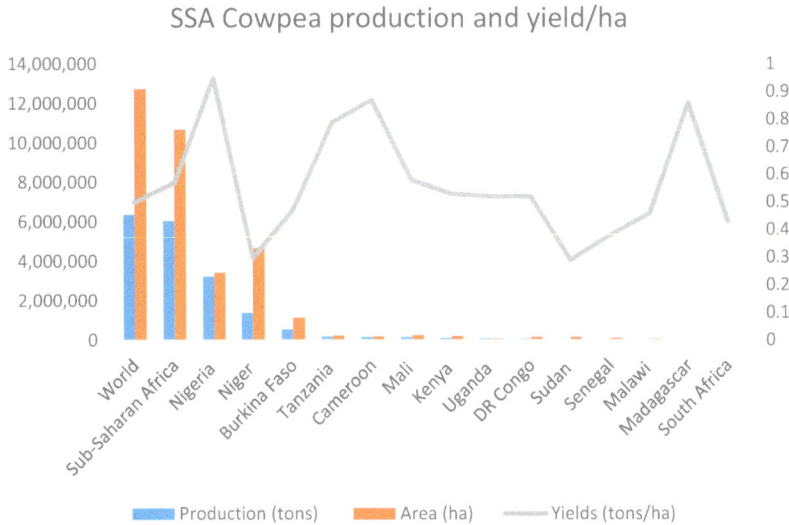

Fig. (9). SSA Cowpea production and yield/ha.

Cowpea yields remain low despite these hopeful developments because of a number of production limitations as well as a lack of adoption of improved varieties and agronomic practices.

Shea Butter

Between western Senegal and northwestern Uganda, in the dry savannas, forests, and parklands of the Sudan zone, shea grows over an estimated 1 million km². With high production zones in Benin, Burkina Faso, Cote D'Ivoire, Ghana, Mali, and Nigeria, it is claimed that [4] Nigeria has the greatest potential for shea nut production. The potential of production capacity is not completely realized in countries where producers, particularly women, and the private sector are underrepresented. The potential of the production capacity is not fully realized since producers, notably women and the private sector in nations where shea trees grow, do not fully participate in the value added sales of the nuts or butter. It is anticipated that this amount will treble once Ghana's shea production potential is fully realized.

The necessity to address the issues in each of the following areas is the main suggestion for the textile and oilseeds industries:

Limited Availability of Better Seeds

The inability of the current seed system to provide smallholder farmers with improved varieties of oil crops and legume seeds is a result of a number of problems. Many farmers are unaware of the investment required because they are used to receiving free seed from non-governmental organizations (NGOs). Creating a Foundation Seed Enterprise dedicated to the production and distribution of foundation/basic seed can help seed companies interested in commercializing improved publicly developed varieties, according to lessons experienced in other parts of the world.

A Lack of Agriculture Equipment

Although yield-improving technologies have been developed during the past three decades, labor-intensive farming methods are still in use, and crop products are still processed manually at home. However, inorganic amendments are rarely accessible to farmers in affordable quantities due to subpar input marketing strategies.

Poor Soil Fertility

However, inorganic amendments are rarely accessible to farmers in affordable quantities due to subpar input marketing strategies.

Input Market Restrictions

Science and technology can aid in increasing agricultural growth in a variety of ways, including improved seeds, fertilizers, crop protection products, and novel agronomic techniques. Inadequate regulation has led to the widespread use of tainted or outdated pesticides.

Market Restrictions

Smallholders have not been able to effectively respond to prospective soybean market opportunities because of a number of structural and institutional barriers that prevent market participation. Processors and traders are constrained by poor grain quality, a lack of supply, and expensive cleaning costs, while market intermediaries struggle with high assembly costs, a high market risk, and cash flow problems. New kinds of market institutions that provide contract creation and enforcement, as well as vertical and horizontal coordination of production and marketing duties, are required to increase smallholder farmers' market access and competitiveness. Farmers' knowledge of and access to new information, their expectations regarding the advantages of new technologies and their local availability, their market access and opportunities, their ability to access credit, and other factors that encourage farmer investment in new technologies have all been shown to be important factors in the diffusion and uptake of research products.

Low Acceptance Rates for New Technologies

Numerous socioeconomic and targeting studies (http://www.icrisat.org/impi-tl-2.htm) show that the production space is still largely occupied by old varieties that were introduced 15 to 20 years ago, indicating that new variety acceptance has been slow and sluggish. Farmers' knowledge of and access to new information, their expectations regarding the advantages of new technologies and their local availability, their market access and opportunities, their ability to access credit, and other factors that encourage farmer investment in new technologies have all been shown to be important factors in the diffusion and uptake of research products.

TEXTILE PRODUCTION CAPACITY AND COMPETITIVE FACTORS

Table **3** below summarizes the textile manufacturing capacity and competitive characteristics of each top country, highlighting each nation's competitive advantages and disadvantages.

In Fig. (**10**) in Appendix 1, changes in volume by country are graphically depicted.

Fig. (**11**) below shows the very fragmented nature of the African Value Chain.

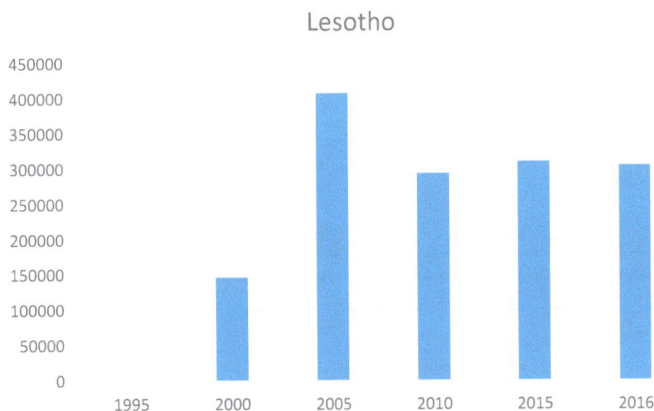

Lesotho

(Fig. 10) contd.....

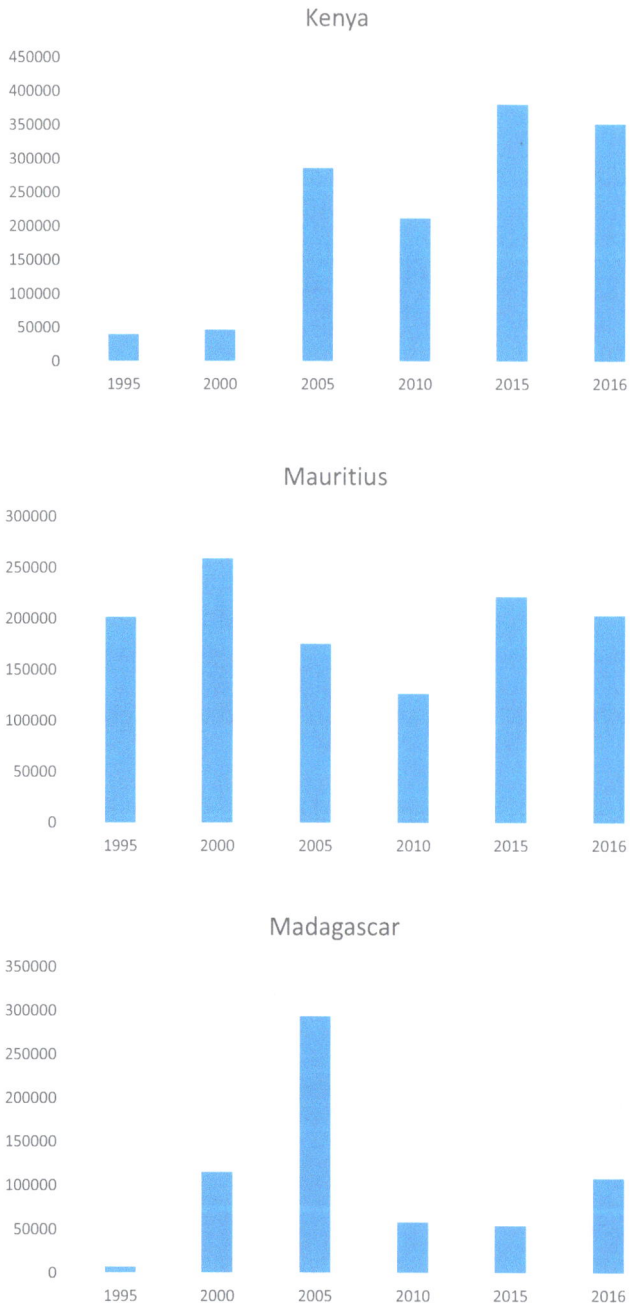

Kenya

Mauritius

Madagascar

(Fig. 10) contd.....

South Africa

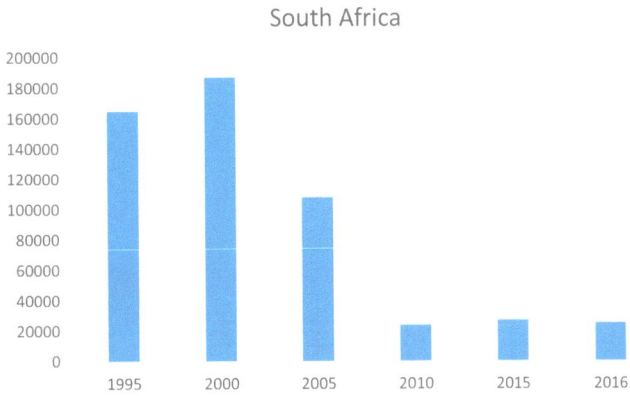

Swaziland

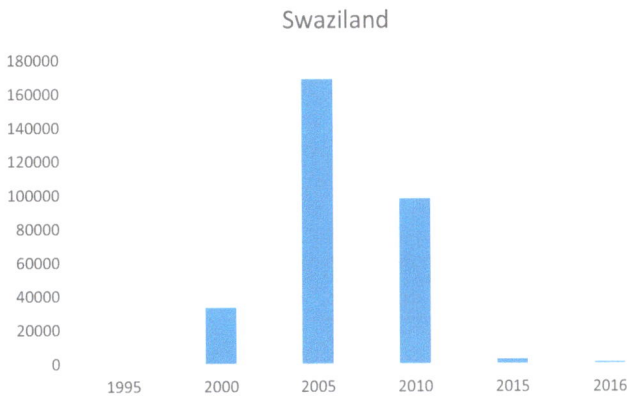

Tanzania

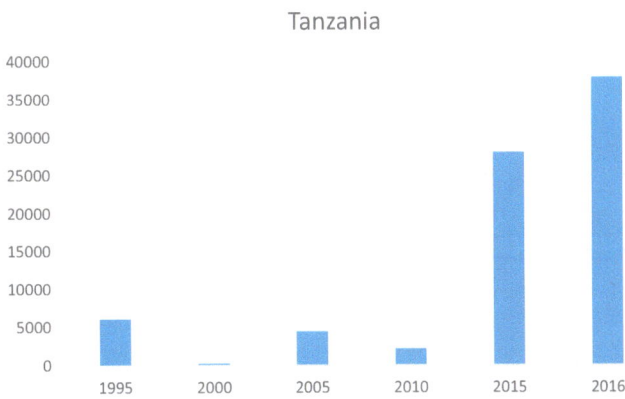

(Fig. 10) contd.....

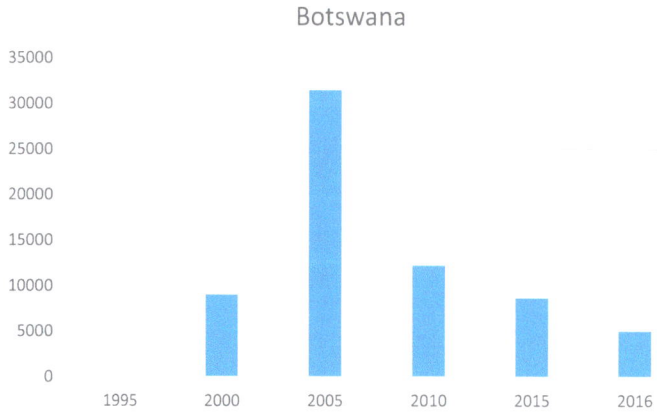

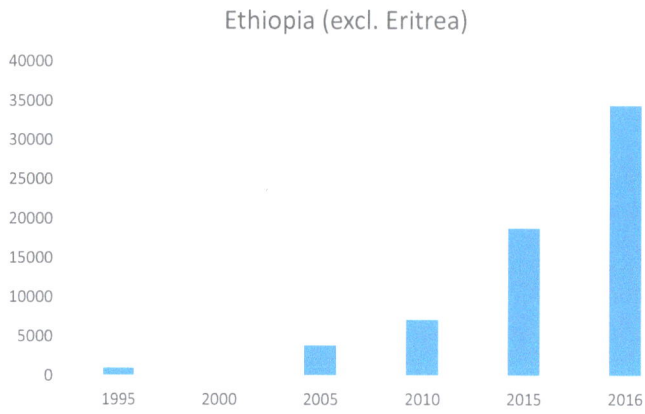

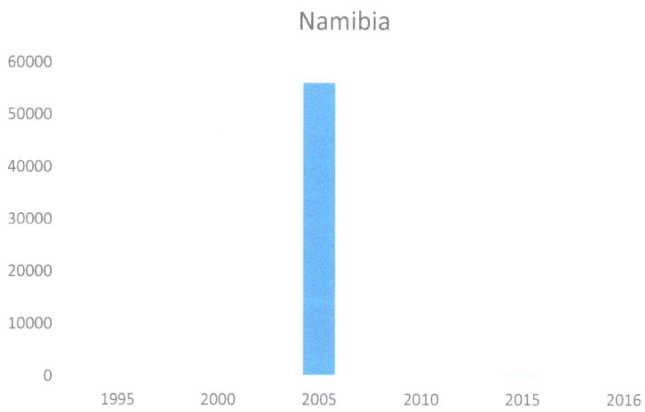

Botswana

Ethiopia (excl. Eritrea)

Namibia

(Fig. 10) contd.....

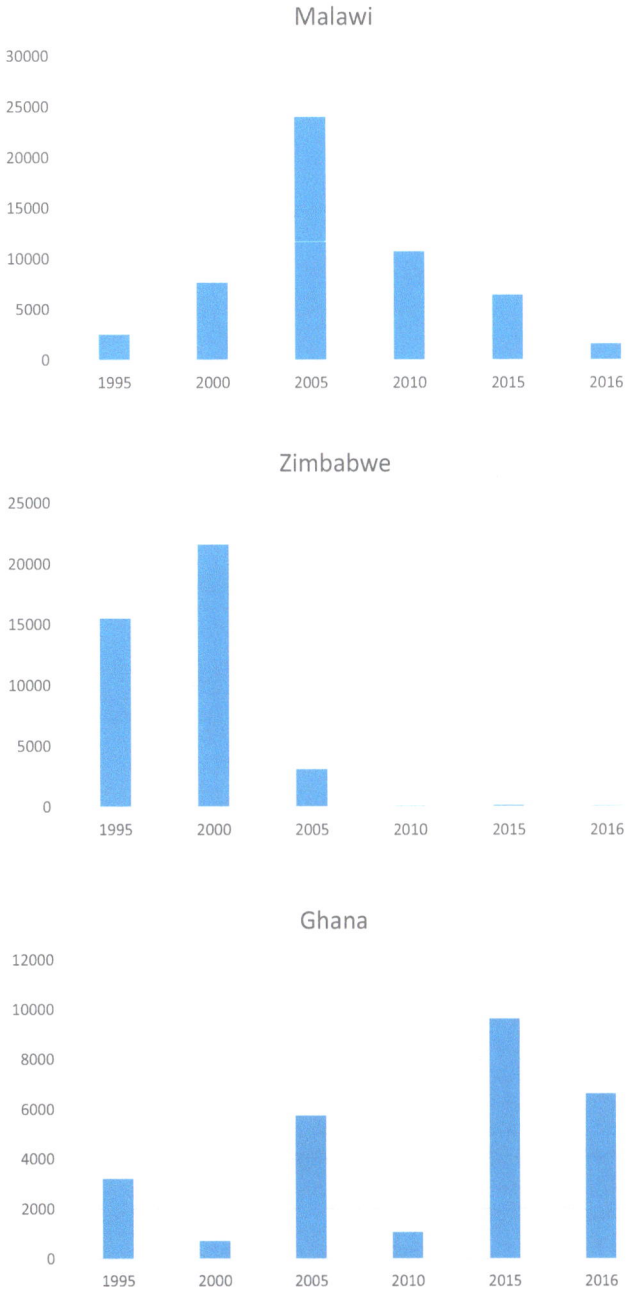

Malawi

Zimbabwe

Ghana

(Fig. 10) contd.....

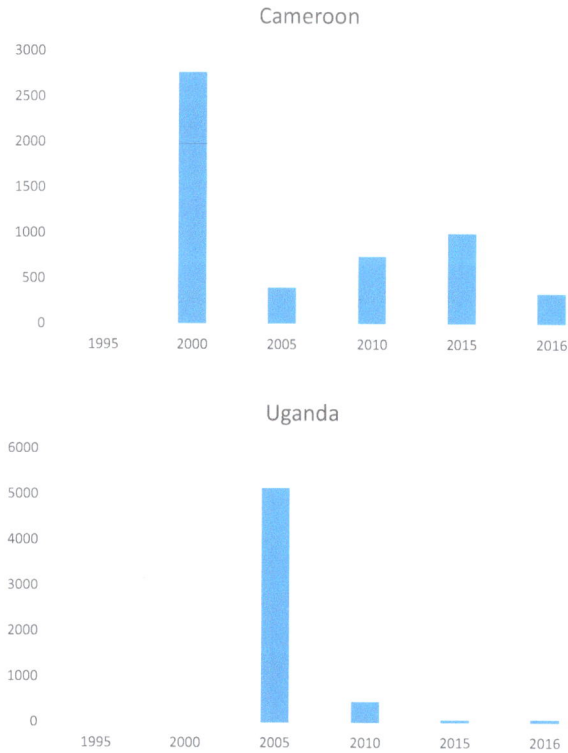

Fig. (10). Changes in the volume of production by country.

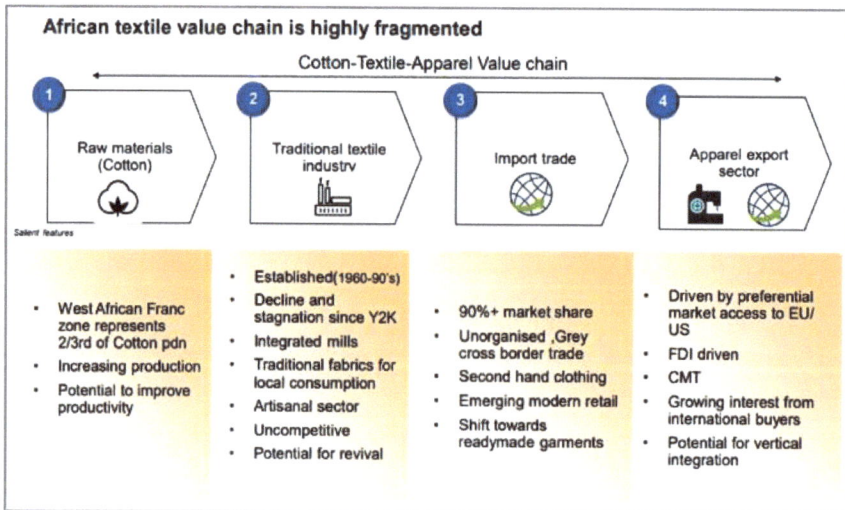

Fig. (11). The African Textile Value Chain.

Table 3. Summary of selected SSA textile and apparel input producers.

Group 1 Countries	Textile and Apparel Input Production	Competitive Factors
Ethiopia	The Ethiopian textile sector includes eight vertically integrated textile mills, along with stand-alone spinning mills for yarn and thread production. Most of the yarn spun in Ethiopia is used in the production of woven cotton fabric. In addition to cotton yarn and woven fabric, Ethiopia's textile sector also produces acrylic yarn, nylon fabric, woolen and waste-cotton blankets, bedsheets, and sewing thread. Ethiopia currently produces cotton and silk yarn for domestic hand-loomed production of niche products, such as home furnishings, for export to the United States, Canada, and Europe.	Competitive advantages: • large potential domestic apparel market • domestic production of raw materials (cotton, silk) • stable political and business environment • access to Ethiopian government-supported investment incentives and financial assistance Competitive disadvantages: • import competition from used clothing • low cotton production; cotton contamination • poor transportation infrastructure • underutilized industrial capacity • outdated machinery and equipment • low labor productivity • lack of skilled labor
Kenya	The Kenyan textile industry has contracted since the 1990s and currently consists of three vertically integrated firms and a few smaller, nonintegrated firms. Kenya's vertically integrated firms produce cotton (including organic) and synthetic yarn, and knitted and woven fabric for use in apparel exported to the United States and the EU. Some yarn and fabric are also sold regionally.	Competitive advantages: • export-oriented apparel industry • relatively skilled labor Competitive disadvantages: • poor roads • high-cost electricity • limited and high cost of financing for new equipment

(Table 3) cont.....

Group 1 Countries	Textile and Apparel Input Production	Competitive Factors
Lesotho	Lesotho has one vertically integrated denim textile mill that spins cotton yarn, dyes the yarn, weaves the fabric, and cuts and sews the finished denim jeans. The mill reportedly produces 10,800 tons of openended ring-spun cotton yarn, and 18 million yards of denim fabric a year for regional apparel manufacturers producing for the export market. Lesotho primarily exports woven fabric to other apparel-producing African countries. The vast majority of Lesotho's apparel exports are to the U.S. market.	Competitive advantages: • export-oriented apparel industry • government investment support for plant acquisitions Competitive disadvantages: • poor water/wastewater and internal transport infrastructure • low labor productivity • high HIV/AIDS prevalence rates • lack of skilled labor
Madagascar	The Malagasy textile industry consists of one large vertically integrated woven textile and apparel firm that consumes most of its own fabric production, two small knit apparel firms that produce their own knit fabric, and another firm that weaves fabric for blankets. The Malagasy apparel sector is geared to supply the U.S. and EU markets.	Competitive advantages: • export-oriented apparel industry • availability of skilled and productive labor • government investment incentives and support Competitive disadvantages: • diminishing supply of domestic cotton • high-cost, unreliable electricity • political instability • high cost of capital • poor road infrastructure

(Table 3) cont.....

Group 1 Countries	Textile and Apparel Input Production	Competitive Factors
Mauritius	The Mauritian industry is concentrated among 10 large textile and apparel groups that collectively account for 75 percent of total textile and apparel exports. The textile and apparel input industry in Mauritius produces yarn and knit fabric mostly for vertical operations, but also for local and regional apparel manufacturers. Mauritius exports textile and apparel inputs to the region and finished apparel primarily to the EU.	Competitive advantages: • export-oriented apparel industry • market linkages with EU apparel buyers • favorable business environment • government support in product and market diversification • relatively modern machinery and equipment • shorter lead times to the region and to some EU customers • availability of skilled labor Competitive disadvantages: • small domestic apparel market • increased labor costs due to labor shortages • long lead times to the United States and to some EU customers • increasing land and energy costs • additional costs associated with geographic isolation
Nigeria	The Nigerian textile industry has contracted since the 1990s and currently consists of 20 or fewer factories. Some larger textile firms are vertically integrated from cotton ginning to spinning, weaving, dyeing, printing, and finishing. The major textile firms produce a variety of products, including polyester staple fiber and filament, yarn, greige cloth, and wax prints. Nigerian printed fabric is sold as loose cloth, rolls, or pieces to the domestic market. Nigerian textile exports are focused on the EU market.	Competitive advantages: • large potential domestic apparel market • history of cotton and integrated textile production • availability of skilled labor Competitive disadvantages: • lack of a developed apparel industry • increased import foreign competition (ethnic cloth and used clothing) • cotton quality issues • poor infrastructure, particularly electricity

(Table 3) cont.....

Group 1 Countries	Textile and Apparel Input Production	Competitive Factors
South Africa	The South African textile sector is relatively large and encompasses the full range of manufacturing operations, including the production of fiber, thread, yarn, knit and woven fabric, nonwovens, trim and accessories, and dyeing and finishing operations. There are currently 11 firms in South Africa producing yarn. Five firms manufacture nonwovens, and reportedly seven firms produce trim, including elastic, buttons, zippers, and similar items. Approximately 16 firms produce woven fabric, while 15 companies produce knit fabric. Of the country's textile producing firms, nine are vertically integrated, manufacturing either yarn through fabric, yarn through finished apparel, or yarn through household textiles. Cotton, wool, mohair, manmade fibers, and natural fibers are used in the domestic textile industry.	Competitive advantages: • large domestic apparel industry • developed infrastructure (transport, power, water) • favorable and stable business environment • large and developed textile industry Competitive disadvantages: • high labor costs • inflexible labor market • lack of skilled labor in the industry • lack of management, marketing, and technical skills • lack of investment • long lead times from order to delivery • highly volatile exchange rate
Swaziland	Swaziland has one integrated textile producer that dyes, spins, and knits cotton fabric (including organic), and then sews the fabric into apparel for export. The firm produces yarn for internal consumption and for export to the region and the EU. Swaziland has an internationally branded zipper producer that supplies local and regional apparel manufacturers.	Competitive advantages: • export-oriented apparel industry • government incentives for foreign direct investment in the textile and apparel industry • reliable electricity supply Competitive disadvantages: • small domestic apparel market • limited amount of local raw materials • labor unrest • high HIV/AIDS prevalence rates

(Table 3) cont.....

Group 1 Countries	Textile and Apparel Input Production	Competitive Factors
Tanzania	The Tanzanian textile sector consists of one independent spinning mill and several integrated firms. The industry spins mostly cotton yarns for both knit and woven fabric. A few fabric mills also blend cotton with polyester or other synthetic fibers; however, all synthetic fibers must be imported. Tanzanian textile mills sell these textiles regionally, or minimally process and print fabric to be sold locally as final products.	Competitive advantages: • availability of good-quality domestic cotton • history of cotton yarn exports to the EU • stable political and economic environment Competitive disadvantages: • lack of a developed apparel industry • unreliable and costly electricity • port delays and congestion • lack of skilled labor • lack of market knowledge • low labor productivity
Zambia	The Zambian textile sector consists of an estimated four knitting/weaving firms and four vertically integrated firms that spin their own yarn for use in finished textile and apparel production. Zambia's textile sector produces primarily 100 percent cotton yarn, along with small quantities of man-mad-fiber yarn, including poly/cotton and acrylic yarn. Most of the yarn produced in Zambia is exported, but a small share is used domestically in the production of woven fabric used to manufacture niche apparel articles such as uniforms and mining work wear, primarily for the local or regional market.	Competitive advantages: • domestic availability of high-quality cotton • open trade regime Competitive disadvantages: • small domestic apparel market • insufficient access to affordable credit • outdated machinery and equipment • lack of skilled labor • low labor productivity • high transportation costs and time • unreliable electricity supply

Kenya, Lesotho, Madagascar, and Ethiopia are the top 4 importers of apparel from SSA to the USA under AGOA. The following vital success elements are necessary for the textile and cotton business in SSA to develop. African nations have five major chances to grow the cotton-textile industry:

1. Restructuring (moving industry from China to other emerging and LDC countries; consolidation and upgrading; pressing economic needs, *etc.)*
2. Endowment (access to plentiful raw materials, the labor force in Africa, land, water, *etc.)*
3. Market access (preferred access to the US and EU markets, RTAs, the AfCFTA, *etc.)*

4. Global projects (Belt & Road Initiative; other international projects, *etc.)*
5. Sustainability (Guidance from the 17 SDGs) (Guidance from the 17 SDGs).

The following areas need policy changes to take advantage of these opportunities:

1. Access to markets
2. Enabling environment
3. Raw material
4. FDI
5. Capacity development

Cheap Chinese imports have wiped out the textile industry of Ghana, Nigeria, Uganda, and other countries. Only four out of thirty textile businesses in Ghana are still operating, according to the Industrial and Commercial Workers Union (ICU). According to the group, the nation used to manufacture yarn for clothing sold both locally and in Sub-Saharan Africa, but that is no longer the case.

CONCLUSION AND RECOMMENDATIONS ON THE COTTON AND TEXTILE INDUSTRY IN SSA

Manufacturing clothing requires a lot of labor, has low startup costs, and easily transferable technologies. As a result, some countries with cheap labor costs, especially those in South and East Asia, have significantly increased their market share during the past forty years.

Following is a summary of the main policy recommendations for LIC governments, industry groups, and apparel companies:

1. Boost internal productivity, talents, and skills to help businesses advance from cut-make-trim (CMT) to full package suppliers.
2. Shorten lead times and boost backward linkages.
3. Upgrade the administrative and physical infrastructure.
4. Increase compliance with labor and environmental laws.
5. Expand end markets into quickly expanding emerging markets.
6. Promote greater regional cohesion.
7. Establish a local garment industry.

BIG DATA ANALYTICS FRAMEWORK MODEL FOR OILSEEDS AND TEXTILE PRODUCTION IN SSA

There are many things to think about, from the idea of a Big Data strategy to the technical tools and capabilities that a corporation should have. The principal benefits of utilizing a Big Data framework are as follows:

1. The Big Data Framework gives firms wishing to start using Big Data or enhance their Big Data capabilities a framework.
2. All organizational structure elements that need to be taken into account in a big data environment are covered by the big data framework.

3. The Big Data Framework is independent of any one provider. By including more nodes, the data storage can be increased.

THE BIG DATA FRAMEWORK'S STRUCTURE

Organizations should take into account the Big Data framework, a systematic approach that consists of six fundamental competencies while developing a Big Data organization. The Big Data Framework depicted in Fig. (12) is shown in the diagram below. Fig. (12) depicts the layout of the Hadoop Architecture.

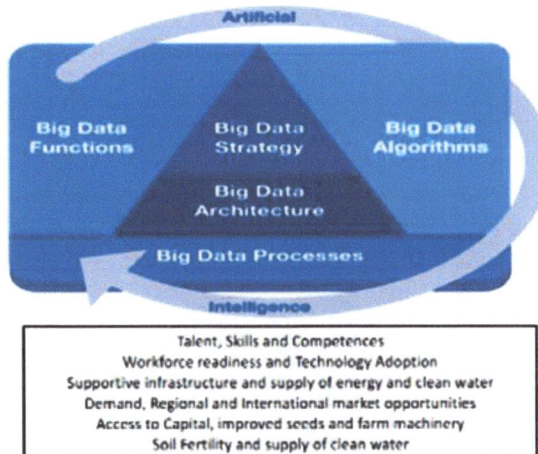

Fig. (12). Big Data Analytics Framework Model for Oilseeds and Textile Production.

Organizations should take into account the Big Data framework, a systematic approach that consists of six fundamental competencies while developing a Big Data organization. When it comes to big data, it's a people business. Businesses will fail even with the most cutting-edge computers and processors if they lack the requisite knowledge and skills. The Big Data Framework works to increase the knowledge of everyone interested in big data result. The modular approach aims to organize Big Data knowledge similarly to how the accompanying certification program does. The Big Data framework as shown in Fig. (**13**) is a thorough method for dealing with big data. It looks at the various factors that companies should take into account while forming a Big Data company.

Hadoop's Architecture: MapReduce Engine

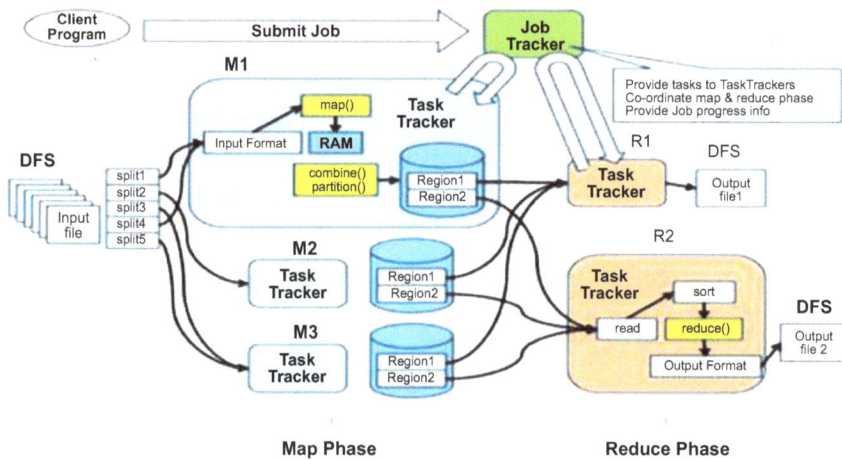

Fig. (13). Hadoop's Architecture.

It looks at the various factors that companies should take into account while forming a Big Data company. Every part of the framework is equally crucial, and companies can only advance if they give each part of the Big Data architecture the same amount of consideration and effort.

The file system, MapReduce engine, and Hadoop Distributed File System are all included in the Hadoop architecture (HDFS). A framework called "map reduce" is intended to process a lot of data concurrently and widely across a group of processing units. A single master and numerous slave nodes make up a Hadoop cluster. DataNode and TaskTracker are on the slave node, whereas Job Tracker, Task Tracker, NameNode, and DataNode are on the master node. The framework aims to simplify interaction with huge data. Through the use of straightforward programming concepts, it enables the distributed processing of big datasets across computer clusters. The Hadoop architecture has established itself in businesses and industries that must work with massive, sensitive data sets that require effective processing [14]. As a result, the Hadoop framework is a framework that permits the processing of huge data sets that are organized into clusters. The

Hadoop architecture consists of a number of modules supported by a sizable ecosystem of technologies and offers a range of services to address big data issues [15]. It consists of Apache projects as well as a number of paid tools and services. Most of the time, these important components are supplemented or supported by tools or solutions. Together, these instruments can offer services, including data absorption, analysis, storage, and maintenance.

CONCLUSION

Big data's importance in the business world is increasing along with technology's advancement. Engagement with government policymakers is required to ensure enough funding for the development of Big Data human capital. Data science courses should be developed for undergraduate and graduate students in order to produce data scientists. In order to give farmers the necessary incentives to invest in better seeds and other complementary inputs to boost productivity and improve quality, one way to achieve this is to persuade commercial seed companies to invest in seed production of publicly developed varieties. This can be done by working with them and other stakeholders to improve coordination along the value chain. The biggest textile companies manufacture a wide range of goods, such as polyester filament, yarn, greige fabric, and wax prints.

The Hadoop platform was created as a framework for big data analytics. The open-source Hadoop platform offers the analytical tools and computing capacity needed to handle such massive data volumes. The Hadoop Distributed File System (HDFS) and the MapReduce parallel processing engine are the two primary parts of Apache Hadoop.

REFERENCES

[1]   J. Sun, and C.K. Reddy, "Big data analytics for healthcare", *Proceedings of the 19th ACM SIGKDD international conference on Knowledge discovery and data mining 2013,* ACM., pp. 1525-1525, 2013.

[2]   T. Kwon, M.I. Chung, R. Gupta, J.C. Baker, J.B. Wallingford, and E.M. Marcotte, "Identifying direct targets of transcription factor Rfx2 that coordinate ciliogenesis and cell movement", *Genom. Data,* vol. 2, pp. 192-194, 2014.
[http://dx.doi.org/10.1016/j.gdata.2014.06.015] [PMID: 25419512]

[3]   G. Kabanda, "Bayesian Network Model for a Zimbabwean Cybersecurity System", In: *Oriental journal of computer science and technology.* vol. 12. Atlantic International University: Honolulu, Hawai, 2020, no. 4, pp. 147-167.
[http://dx.doi.org/10.13005/ojcst12.04.02]

[4]   "OECD-FAO Agricultural Outlook", In: *OECD Agriculture statistics* OECD iLibrary, 2016.
[http://dx.doi.org/10.1787/888933381520]

[5]   E. Alpaydin, *Machine learning. the mit press essential knowledge series.* MIT Press: London, England, 2016.

[6]   T.C. Truong, Q.B. Diep, and I. Zelinka, "Diep Qb, and Zelinka I. (2020). Artificial intelligence in the cyber domain: Offense and defense", *Symmetry (Basel),* vol. 12, no. 3, p. 410, 2020.

[http://dx.doi.org/10.3390/sym12030410]

[7]   K. NAPANDA, H. Shah, and L. Kurup, "Artificial Intelligence Techniques for Network Intrusion Detection", *International Journal of Engineering Research & Technology (IJERT),* vol. 4, no. 11, 2015.

[8]   P. Adriaans, and D. Zantinge, *Data mining.* Addison-Wesley Longman Publishing Co., Inc., 1997.

[9]   "International Telecommunication Union (ITU)", *Global Security Report.,* 2017.

[10]   Shambhoo Kumar, "Predicting and Accessing Security Features into Component-Based Software Development: A Critical Survey", In: *Advances in Intelligent Systems and Computing* vol. 731. Springer, 2019, pp. 287-294.

[11]   K. Bakshi, "Considerations for big data: Architecture and approach", In: *2012 IEEE aerospace conference.* IEEE, pp. 1-7, Big Sky, MT, USA, 2012. [http://dx.doi.org/10.1109/AERO.2012.6187357]

[12]   R. Saunders, J.W. Campbell, and C. Freese, "Change and employee behavior", *Leadersh. Organ. Dev. J.,* vol. 19, no. 3, pp. 157-163, 2009.

[13]   J. Dean, and S. Ghemawat, "MapReduce", *Commun. ACM,* vol. 51, no. 1, pp. 107-113, 2008. [http://dx.doi.org/10.1145/1327452.1327492]

[14]   B. Chowdhury, T. Rabl, P. Saadatpanah, J. Du, and H.A. Jacobsen, "A bigbench implementation in the hadoop ecosystem", In: *Advancing big data benchmarks.* Springer: Cham, 2013, pp. 3-18.

CHAPTER 9

A Design of Lighting and Cooling System for Museum and Heritage Sites

Amrapali Nimsarkar[1], **Piyush Kokate**[2,*], **Mamta Tembhare**[3] and **Harikumar Naidu**[1]

[1] *Department of Electrical Engineering, GHRCE, Nagpur, India*

[2] *Energy & Resource management Division, CSIR-NEERI, Nagpur, India*

[3] *Waste Processing Division, CSIR-NEERI, Nagpur, India*

Abstract: Museums, buildings and heritage sites need artificial light at night time in darker places. At many museums, old lighting is used to illuminate the central gallery section or paintings as well. There are old lighting, including Metal Halide, Incandescent Lamp, Sodium Vapor, CFL, *etc.,* that consume more electricity and produce heat in the indoor environment, causing damage to the artwork, walls, and paintings. No standard guidelines or methodologies have been adopted by our country for lighting at the museums and archeological sites to maintain an elegant look during the day-night time. It is intended to expand in this arena due to a lack of knowledge in the field of lighting at museums as well as at heritage sites.

This paper discusses the correlation of lumen and temperature on different materials by using an LED lighting module with fiber optic cable. ANOVA method was used to correlate the dependent parameters like lumen and temperature concerning a change in distance and time on a material. We have used a lighting module that helps to prevent damage to the objects and emits negligible heat in the environment so that visitors can easily visualize the objects with proper lux level.

Keywords: LED, Museum lighting, Material, Optical fiber.

INTRODUCTION

The museum and heritage sites are the main sources of attraction for foreign visitors in Asian countries [1]. UNESCO has already identified fourteen primary factors as a threat to heritage sites which includes temperature, relative humidity, light, dust, wind, paste, water, microorganisms, *etc.,* under the local conditions

---

* **Corresponding author Piyush Kokate:** Energy & Resource management Division, CSIR-NEERI, Nagpur, India;
E-mail: pa_kokate@neeri.res.in

Hemachandran K., Raul V. Rodriguez, Umashankar Subramaniam & Valentina Emilia Balas (Eds.)

and physical fabrics category [2]. Most of the monuments and heritage sites are illuminated using traditional lighting technologies like CFL, LED, *etc.,* which generate heat [3].

Recent scientific studies reveal that the deterioration of sculptures, wall paintings, building materials and murals can be possible due to an increase in the indoor environmental temperature, insects and lux levels [4]. Museum sites are made up of different types of materials such as stones, wood, rock, sand, *etc* [5, 6]. Physical agents such as temperature, humidity, water, and the paste can deteriorate the structural health of buildings at a slow speed [7]. The indoor and outdoor temperature environment varies from 5°C - 10°C during temperature in summer and winter seasons. The insects, infestation of bats, get attracted towards temperature zone locations. The nonsystematic visits of tourists and the CO_2 emissions from visitors also contribute to the change of indoor temperate; additionally, traditional lighting devices contribute significantly to the change in the indoor temperature. Therefore, the cooling system enhanced by the replacement of traditional lighting with LED is beneficial and energy-efficient compared to High-Intensity Discharge (HID) incandescent lights and fluorescent lamps [8]. The shorter wavelength of light damages the artifacts because it has higher photon energies that are more damaging to the exhibits. Therefore, blue lights are restricted for human exposure [9]. LED has 6500k, which is cooler than other lamps, such as an incandescent lamp, which has 2800k and a fluorescent light with 2700- 6000k, *etc* [10]. According to Author Kohtaro, the LED does not provide UV radiation and IR radiation, compared with the other lighting devices. Also, LED has a high potential to replace the traditional lighting devices with a wide color range and dimming option. It produces cool light, mostly appropriate for caves, because it generates a soothing environment for visitors. In fiber-optic LED, the light rays are passed through one end of a fiber optic cable. The light gets dissipates through the core edge of the tube. At another end of the fiber optic cable, the desired output illumination can be obtained using a filter arrangement. End emitting fiber and edge-emitting fibers are divided as per the emission capacity through the core and cladding area. This glass fiber transfers light radiations uniquely, so it can be used for caves, museums, underground fortification, *etc.* The first part of this article discussed the national protocols and literature available on the lighting & illumination related studies. Then design of fiber optic LED with its circuits and experimental results on material are discussed in a systematic manner. Then the final part discussed how the fiber optic LED can be a prominent option for focused illumination [11-13].

METHODOLOGY

A preliminary experimentation work was conducted at lab level at CSIR-NEERI to study the types of stones and materials used at such sites. It was found that most of the building structures are made up of stone, granite, bricks, tiles and concrete [14]. It was difficult to carry out the study directly at the museum or heritage sites due to restrictions and permission issues. Therefore, a fiber optic-based LED lighting device was designed and tested in an artificial chamber to study the impact on materials. Lux meter and temperature sensors were used to study the lumen and temperature with varying time and distance from the source [15]. The light was reflected on the material through LMS at a variable constant distance of 0.5m, 1m, 1.5m, and 2m with variable time duration of 10 min, 20 min, 30 min, and 40 min. Lux meter 101 A comprised of 1 X 0.9V of the power source with 0-200000 LUX range was used. Temperature sensor (EXTECH) [16] and temperature indicator (TM25) models were used to measure the change in temperature of material inside the acrylic chamber. Further, this illuminated light from the power LED is passed through the optical fiber cable. A hollow square chamber of dimension 30 cm x 30 cm was created for laboratory study, which is shown in Fig. (1).

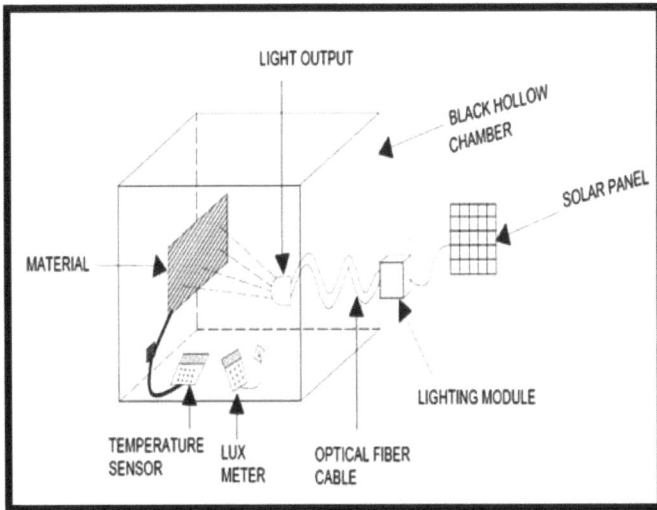

Fig. (1). Testing Method of Lighting Module.

Pre-Scanning of LMS

Ocean Optics USB 2000 spectrometer was used to capture the wavelength of different lighting devices. Acrylic material was used to construct the hollow chamber, and the hardware module designed is shown in Fig. (2).

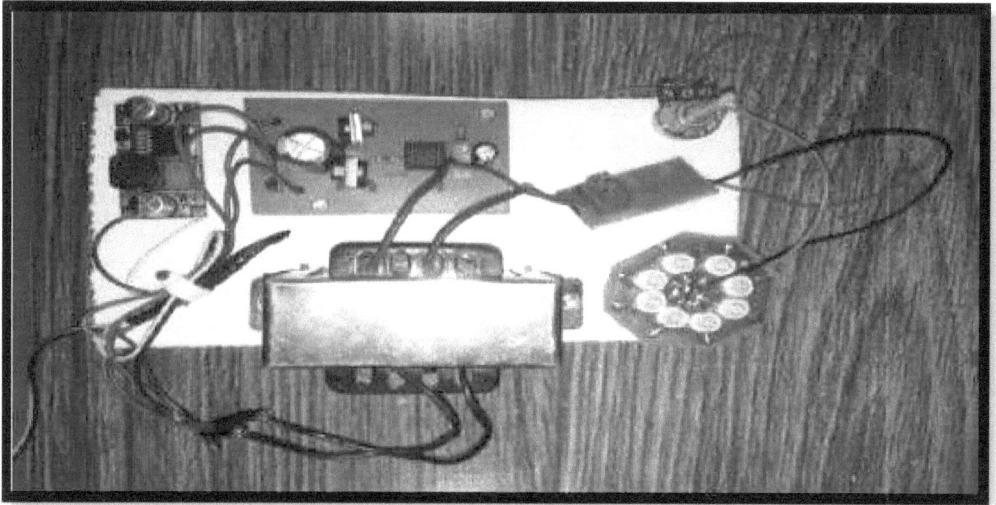

Fig. (2). Hardware Module.

Results and Discussion: Lighting Module Design

A separate Lighting Module System (LMS) was developed, which was mounted in the hollow deep part of the chamber. LMS comprised of a solar panel (5W), buck converter (4-35V), battery voltage (7.8V), inverter circuit (AC 120V), potential meter (10V), rectifier IC, power LED (1W) and fiber optic cable [17]. The Power of 10V is generated by means of a solar panel. The power is transferred to Buck Converter (BC) to reduce the input voltage to 8V. The output supply from Buck Converter is connected through the Zener diode, battery and inverter circuit. Also, an 8.2V of Zener diode is used to protect against the overcharging of the battery. The power supply of 7.2 V is given to the inverter circuit through the battery to convert DC to AC. A transformer is used to boost the high voltage of 120 AC and the rectifier IC is used to convert 120 AC to DC. Intensity is controlled by using a 10V potentiometer. The values for the wavelength of different lights of varying wavelength are shown in Fig. (3). The wavelength scan of the LED lamp is drawn with lumen vs. wavelength, which had depicted the highest wavelength at 450 nm. The wavelength of LED ranges between 400nm to 600 nm. It is the visible limit for the human being.

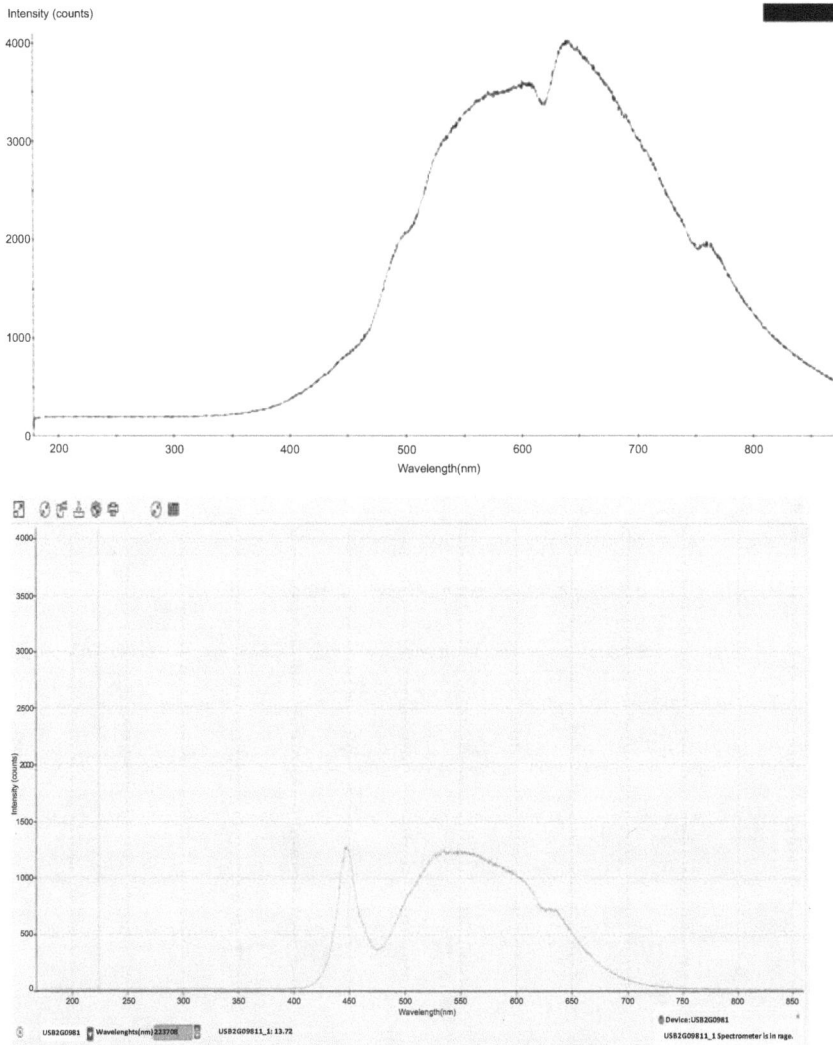

Fig. (3). Testing Method of Lighting Module.

Cooling Effect Study by Temperature Monitoring *Vs* Time and Distance

The designed module was used to estimate the change in temperature variation with respect to distance and time. To analyse the maximum variation in temperature on multiple surfaces, a special condition was created by using a black box. The light module was installed in a black box which has comprised of low loss in light emission. The light module was installed in the black box to test the different building materials and their heating effects. The experiment was carried out for seven days on different materials with continuous monitoring. ANOVA method was used to co-relate the dependent parameters like lumen and temperature with respect to change in distance and time on a material [18]. The LED and Incandescent light is used for evaluating and comparative analysis with variable distances such as 5 cm, 10 cm, and 15 cm. The result is displayed in Table **1** for incandescent light and Table **2** for LED light. The graphical representation of material and temperature effects is shown in Fig. (**4**), and the material *vs.* lumen is shown in Fig. (**5**).

Table 1. Temperature and Lumen Value for Incandescent Lamp.

Material	Distance(cm)	Temperature(°C)	Lumen Value(Lux)
Tiles	5,10,15	31.3	104
Bricks	5,10,15	33.2	92
Concrete	5,10,15	32.9	96
Stone	5,10,15	33.7	98
Granite	5,10,15	32.5	97

Table 2. Temperature and Lumen Value for Incandescent Lamp.

Material	Distance(cm)	Temperature(°C)	Lumen Value(Lux)
Tiles	5,10,15	(30.1,30.2,30.3)=30.2	(126,111,94)=110
Bricks	5,10,15	(30.8,30.9,31.0)=30.9	(112,105,82)=99
Concrete	5,10,15	(30.0,31.2,31.4)=31.0	(123,114,85)=107
Stone	5,10,15	(30.8,31.4,31.6)=31.2	(124,110,91)=108
Granite	5,10,15	(30.5,30.6,30.7)=30.6	(121,108,90)=106

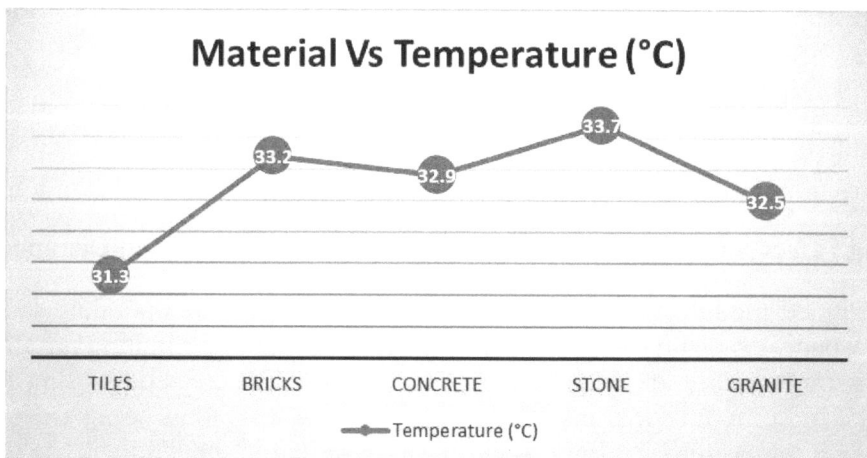

Fig. (4). Testing Material *vs.* Temperature for Incandescent Lamp.

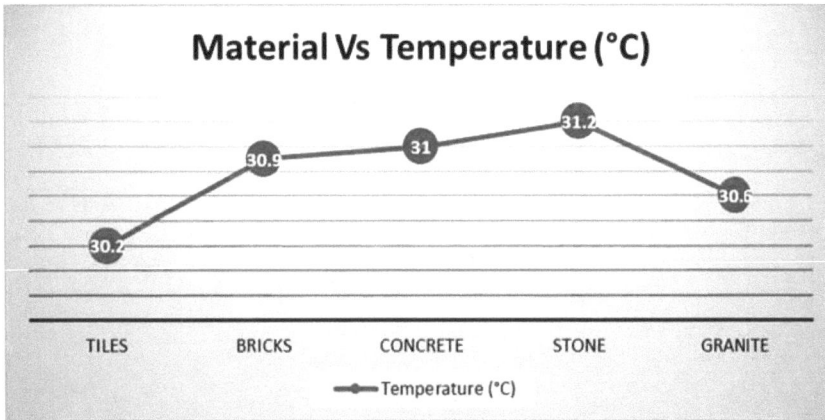

Fig. (5). Material *vs.* Temperature of LED.

The light transmitted inside the heritage building creates a particular value of temperature. To view the object's visibility, it is necessary that the artificial light effect at night time is observed. Museums, caves and Heritage buildings use different lights with varying degrees of temperature. As per the literature study, the warmer temperature contributes to the growth of microbiological factors such as algae, fungi, bugs, and insects, which start breeding inside the cave. These result in the excretion of chemical material on the paintings and walls. This is the major cause of the deterioration of heritage walls, Murals, artifices and paintings. This experimentation used light with a varying lumen which is measured with the help of a lux meter simultaneously. The temperature value was noted by using a temperature sensor. The lumen value and temperature value were measured between the distance of 5.0 cm to 15cm. The light output was kept at a distance of 5cm from the measurement point. At a certain lumen value, the temperature was recorded by using a temperature sensor. The result shows that temperature increases gradually when the lumen value increases. The light output was kept at a distance of 10cm from the measurement point. At a certain lumen value, the temperature was recorded by using the temperature sensor. The result has shown that temperature increases gradually when the lumen value increases. Similarly, from a distance of 15cm, the temperature value was slowly growing. The graphical representation of lumen *vs.* temperature is shown in Fig. (**6**).

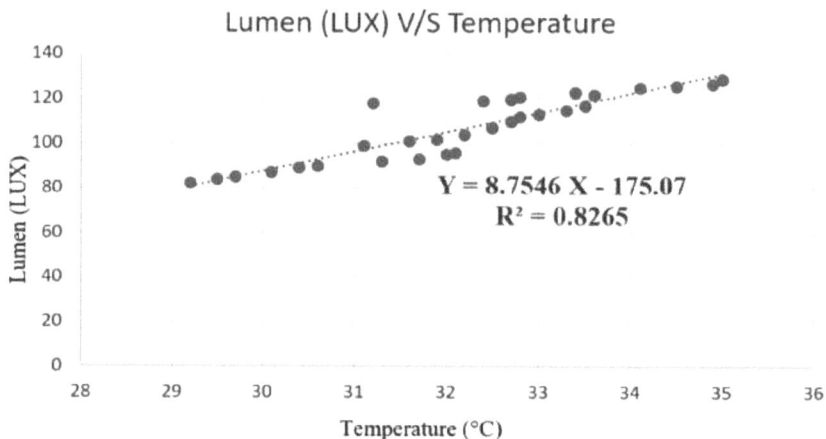

Fig. (6). Lumen *vs.* Temperature of LED.

The statistical analysis ANOVA (Analysis of Variance) shows the correlation between the lumen and temperature, which is found to be 82%, and the regression equation used is:

$Y = -275.43 + 11.55x$ …….. (1)

$R^2 = 0.8265$……. (2)

Proposed Cooling Effect Enhancement System Design

The LMS design can be installed inside any building/museum site as per the availability of a power source. The light radiation will travel through the fiber optic shown in Fig. (7).

Then the lenses at the end of the cable will allow only light radiation to emit. The source of this light will be kept outside the structure to protect the location from thermal effects and increase the cooling aspect. Such designs will help the visitors to visualize the objects due to the adequacy of lux level emitted through optical fiber cable with minimum thermal effects. Lower lux level and direction of light will protect the site from glare formation. The illuminance and reflection factor will not be uniform inside the indoor environment.

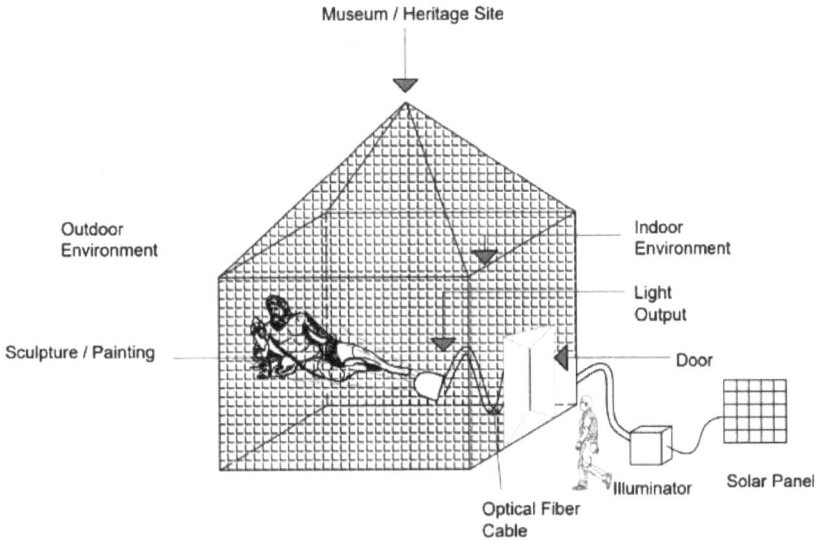

Fig. (7). Design of a Museum of Heritage Sight.

CONCLUDING REMARKS

The cooling effect design was experimentally tested and analyzed on a fiber optic-based LED module, which is useful for the indoor environment at museums and heritage sites. The effect on the change in temperature due to such lighting effect shows major protection from indoor degradable environments. Therefore, the proposed devices are only the prominent option for the low lumen requirement areas around the heritage sites. The experimental result also reveals that the stone structure is more prone to change in temperature due to illumination, hence optical device proposed is only the best solution to be adopted in the future. Archeological Survey of India (ASI) needs to replace the traditional lighting devices with Modern lights (LED), which will reduce energy consumption and maintain balance in the lighting and cooling sector.

REFERENCES

[1]   K. Hosseini, A. Stefaniec, and S.P. Hosseini, "World Heritage Sites in developing countries: Assessing impacts and handling complexities toward sustainable tourism", *J. Destin. Marketing Manag.,* vol. 20, p. 100616, 2021.
[http://dx.doi.org/10.1016/j.jdmm.2021.100616]

[2]   "UNESCO World Heritage Centre - List of factors affecting the properties", Available at:https://whc.unesco.org/en/factors/(accessed on: 2021).

[3]   C.D. Galaţanu, "Lighting systems calculations for heritage buildings, IOP Conference series", *Mater. Sci. Eng.,* vol. 586, p. 012016, 2019.

[4]   S. Singh, S. Dhyani, P. Kokate, S. Chakraborty, and S. Nimsadkar, "Deterioration of World Heritage Cave Monument of Ajanta, India: Insights to Important Biological Agents and Environment Friendly Solutions", *Heritage,* vol. 2, no. 3, pp. 2545-2554, 2019.
[http://dx.doi.org/10.3390/heritage2030156]

[5]   O.P. Agrawal, S. Dhawan, K.L. Garg, F. Shaheen, N. Pathak, and A. Misra, "Study of biodeterioration of the Ajanta wall paintings", *International Biodeterioration,* vol. 24, no. 2, pp. 121-129, 1988.
[http://dx.doi.org/10.1016/0265-3036(88)90054-1]

[6]   O. P. Agrawal, "Conservation problems of Ajanta wall paintings", *Studies in Conservation,* vol. 31, no. 1, pp. 86-89, 2013.
[http://dx.doi.org/10.1179/sic.1986.31.Supplement-1.86]

[7]   G. Bharti, "Ajanta caves: Deterioration and Conservation Problems (A Case Study)", In: *Int. J. Sci. Res. Publ.* vol. 3. University of Lucknow: Uttar Pradesh, India, no. 11, 2013.

[8]   "The Top Five Global Lighting Technologies", Available at:https://docplayer.net/42858837-The- top-five- global-lighting-technologies.html(accessed on: 2021).

[9]   D. Saunders, and J. Kirby, "Wavelength-dependent fading of artists' pigments", *Stud. Conserv.,* vol. 39, no. sup2, pp. 190-194, 1994.
[http://dx.doi.org/10.1179/sic.1994.39.Supplement-2.190]

[10]   M.S. Islam, R. Dangol, M. Hyvärinen, P. Bhusal, M. Puolakka, and L. Halonen, "User preferences for LED lighting in terms of light spectrum", *Light. Res. Technol.,* vol. 45, no. 6, pp. 641-665, 2013.
[http://dx.doi.org/10.1177/1477153513475913]

[11]   M. Górczewska, "Some aspects of architectural lighting of historical buildings", *Lighting in Engineering, Architecture and the Environment,* vol. 121, pp. 107-116, 2011.
[http://dx.doi.org/10.2495/LIGHT110101]

[12]   H.F.O. Mueller, "Energy efficient museum buildings", *Renew. Energy,* vol. 49, pp. 232-236, 2013.
[http://dx.doi.org/10.1016/j.renene.2012.01.025]

[13]   R.E. Caraka, S. Shohaimi, I.D. Kurniawan, R. Herliansyah, A. Budiarto, S.P. Sari, and B. Pardamean, "Ecological show cave and wild cave: negative binomial gllvm's arthropod community modelling", *Procedia Comput. Sci.,* vol. 135, pp. 377-384, 2018.
[http://dx.doi.org/10.1016/j.procs.2018.08.188]

[14]   A. Bulakh, P. Härmä, E. Panova, and O. Selonen, "Rapakivi granite in the architecture of St Petersburg: a potential Global Heritage Stone from Finland and Russia", *Spec. Publ. Geol. Soc. Lond.,* vol. 486, no. 1, pp. 67-76, 2020.
[http://dx.doi.org/10.1144/SP486-2018-5]

[15]   P. Lassandro, T. Cosola, and A. Tundo, "School Building Heritage: Energy Efficiency, Thermal and Lighting Comfort Evaluation Via Virtual Tour", *Energy Procedia,* vol. 78, pp. 3168-3173, 2015.
[http://dx.doi.org/10.1016/j.egypro.2015.11.775]

[16]   Copley Herbert, G. Andrew, H. Emily, D. Edward, and Baddour Natalie, "Temperature and measurement changes over time for F-Scan sensors", *In 2013 IEEE International Symposium on Medical Measurements and Applications (MeMeA).* IEEE, pp. 265-267, Gatineau, QC, Canada, 2013.
[http://dx.doi.org/10.1109/MeMeA.2013.6549749]

[17]   C. Balocco, and G. Volante, "A Method for Sustainable Lighting, Preventive Conservation, Energy Design and Technology—Lighting a Historical Church Converted into a University Library", *Sustain,* vol. 11, p. 3145, 2019.
[http://dx.doi.org/10.3390/su11113145]

[18]   R.G. O'brien, "A general ANOVA method for robust tests of additive models for variances", *J. Am. Stat. Assoc.,* vol. 74, no. 368, pp. 877-880, 1979.
[http://dx.doi.org/10.1080/01621459.1979.10481047]

CHAPTER 10

Predict Network Intruder Using Machine Learning Model and Classification

Chithik Raja[1,*], **Hemachandran K.**[2], **V. Devarajan**[1] and **K. Jarina Begum**[3]

[1] *University of Technology and Applied Sciences Salalah, Salalah, Sultanate of Oman*

[2] *Department of Artificial Intelligence, School of Business, Woxsen University, Hyderabad, India*

[3] *Jazan University, Jazan, Kingdom of Saudi Arabia*

Abstract: The massive number of sensors deployed in IoT generates humongous volumes of data for a broad range of applications such as smart home, smart healthcare, smart manufacturing, smart transportation, smart grid, smart agriculture *etc*. Analyzing such data in order to facilitate enhanced decision making and increase productivity and accuracy is a critical process for businesses and life improving paradigm. Machine Learning would play a vital role in creating smarter techniques to predict the intruder from the dataset. It has shown remarkable results in different fields, including Network security, image recognition, information retrieval, speech recognition, natural language processing, indoor localization, physiological and psychological state detection, *etc*. In this regard, intrusion detection is becoming a research focus in the field of information security. In our experiment, we used the CICIDS2017 data set to predict the Network Intruder. The Canadian Institute of Cyber Security released the data set CICIDS-2017, which consists of eight separate files and includes five days' worth of normal cum abnormal network packet data. The goal of this research is to examine relevant and significant elements of large network packets in order to increase network packet attack detection accuracy and reduce execution time. We choose important and meaningful features by applying Information Gain, ranking and grouping features based on little weight values on the CICIDS-2017 dataset; and then use Random Forest (RF), Random Tree (RT), Naive Bayes (NB), Bayes Net (BN), and J48 classifier algorithms. The findings of the experiment reveal that the amount of relevant and significant features produced by Information Gain has a substantial impact on improving detection accuracy and execution time. The Random Forest method, for example, has the best accuracy with 0.14% of negative results when using 22 relevant selected features, whereas the Random Tree classifier algorithm has a higher accuracy with 0.13% of negative results when using 52 relevant selected features but takes a longer execution time.

Keywords: Accuracy, CICIDS2017, Classification, Execution time, Information Gain, Model Prediction, Recent Data Set.

---

* **Corresponding author Chithik Raja:** University of Technology and Applied Sciences Salalah, Salalah, Sultanate of Oman; Tel: 94758648; E-mail: chithik43@gmail.com

INTRODUCTION

The rapid advancements in information and communication technologies around the world present a significant challenge for network engineers.

The detection of evil behaviors in a host that later spread to other hosts over a network is a major concern for today's network engineers and researchers. Intrusion is a term used to describe an untrustworthy program that is forced to participate in this calamity. Intrusion is defined as unauthorized access to the system or network resources. Intrusion Detection Systems (IDS) serve a crucial role in reducing such activities. The majority of IDSs use anomaly detection or misuse detection mechanisms. In companies, misuse detection mechanisms are common for creating successful commercial IDS, whereas heuristic analysis mechanisms are confined to create successful commercial IDS, whereas anomaly detection is reserved for academic research and development [1]. However, in order to identify future assaults, an IDS requires existing data. That is why IDSs were previously trained on a useful dataset.

Numerous studies have been published which use techniques for selecting features to boost anomaly detection effectively. The Network Security Laboratory-Knowledge Discovery and Data Mining (NSL-KDD) is an updated and enhanced version of the KDD Cup 99 dataset used in the majority of the studies. To increase classification rate, various techniques and metrics have been suggested, including Chi-Square, Information Gain, Correlation Based with Naive Bayes and Decision Tree Classifier, Support Vector Machine (SVM), and Random Forest [2]. Those strategies, however, were not put to the test on a large dataset with a significant number of features. As indicated in [3], data with large datasets can alter the learning model, which tends to overt, lowering efficiencies, increasing memory footprint, and increasing analytic processing cost. In reality, the computational time is rarely considered by researchers, particularly in anomaly detection have regularly used Information Gain to examine significant and relevant aspects.

The Information Gain is utilized to minimize the dimension of the data by picking the most relevant characteristics with feature weight calculation [4, 5].The detecting system's performance may be improved by removing irrelevant features. Many research studies use Information Gain to examine a dataset with few attributes. In this analysis, the CICIDS-2017 dataset with its more intricate properties is used. A large variety of attributes and a large volume of packets are available in the CICIDS-2017 dataset, which can be utilized to identify anomalies.

The Information Gain feature selection technique has been employed in earlier studies that used the CICIDS-2017 dataset, but those studies were unable to explain how or why the scoring value used alone during feature selection is derived [3]. The distinct score value is given by every researcher. The capacity of Information Gain to choose significant characteristics for network packet classification, particularly for data with a huge dataset, is investigated and analyzed [6 - 8]. The features are divided into groups by their minimal score values. Our experiment selects pertinent and significant features by using ranking, grouping and Information Gain features based on a few weighting factors, and implementing the Random Forest (RF), Naive Bayes (NB), Bayes Net, and Random Tree classifier algorithms in experiments on the CICIDS-2017 dataset. The experiment's results show that enhancing detection accuracy and execution time is significantly impacted by the quantity of meaningful and relevant characteristics created by Information Gain.

RELEVANT RESEARCH

The problem of feature selection in relation to network attack detection has been studied by choosing the most important features [9], which uses a combination of filtered-based and wrapper-based algorithms to evaluate the features of big network packets. The method creates ten significant features, has a 0.2% of the unidentifiable rate, and a 0.34 percent false alarm rate. A supervised filtered-based features selection method called Flexible Mutual Information Feature Selection (FMIFS) [10] has been presented. Compared to earlier techniques, the strategy improves the Least-Squares Support-Vector Machines (LS-SVM) IDS's computational efficiency and dependability.

In order to give weight to every feature [11], a study proposes a feature identification approach that blends filtered- and wrapper-based methods with a clustering algorithm. The proposed approach can find features that could really improve attack detection accuracy. Literature [12] describe a tree-seed algorithm (TSA) for extracting useful features. The proposed method reduces the dimension of data by removing redundant characteristics and increasing the precision of the K-Nearest Neighbor (KNN) classifier. The proposed strategy provides 16 significant characteristics with a classification accuracy of 99.92 percent using the C4.5 Machine Learning method and a Discrete Differential Evolution (DDE) mechanism. The FACO algorithm, which combines the Ant-Colony Optimization technique and feature selection, is used by Peng *et al.* The described method can result in features that improve the classification algorithm's precision. Finally, researchers [2] suggest an IDS called FWP-SVM-GA based on the genetic algorithm and SVM. This method reduces the number of false positives while increasing the detection rate, accuracy, true positive rate (TPR), and SVM testing set (FPR).

After a review of past research, we discovered that by deleting irrelevant and superfluous features from classification systems, feature selection might improve their efficiency. The accuracy of detection can be increased even with a small number of selected features. The KDD CUP 99 dataset comprises just 41 attributes. It contains more testing and training datasets used by investigators. Using a huge dataset is still challenging as a result, the efficiency of the suggested methodologies has not been verified on a larger dimension dataset (with more features and events). Table **1** summarizes feature selection research activities in the intrusion detection industry over the previous five years. To increase the performance of AdaBoost-based IDS on the CICIDS-2017 Dataset, researchers [2] combine the Synthetic-Minority-Oversampling-Technique (SMOTE), Principal Component Analysis (PCA), and Ensemble Feature Selection (EFS). In terms of accuracy, precision, recall, and F1 Score, the researchers state that the methodology beats the SVM-based method.

Table 1. Summary of related study.

References	Feature Selection Algorithm	Dataset	No. of Features	Result
[21]	Filtered based Information-Gain enhanced with Baysian Wrapper with C4.5	KDD Cup99	41	Best detection rate with 0.3% False positive rate
[20]	Flexible Mutual Information Feature Selection (FMIFS)	KDD Cup99,NSL-KDD and Kiyota2006+	41,41,24	Improved the accuracy and low computational cost in comparison to current best practices.
[19]	Multi measured Multi Weight Feature Selection (Filtered Based and Wrapper based)	KDD Cup99	41	Good accuracy rate with less detection time
[18]	DDE,C4.5 ML algorithm	NSL-KDD	41	16 relevant features with 99.92% of classification accuracy
[17]	Tree-Seed+ algorithm with KNN Classifier	KDD Cup99	41	While remove the redundant increase the effectiveness and precision of network intrusion detection
[18]	Ant Colony	KDD Cup99	41	Improve Classification accuracy with 98%

(Table 1) cont.....

[19]	Feature selection, Weight and Parameter Optimization (FWP) to support Genetic Algorithm and SVM	KDD Cup99	41	Boost the rate of detection, accuracy, and true positives. Reduce the SVM training time and false positive rate.
[20, 21]	PCA,SMOTE	CICIDS-2017	78	Improves IDS performance on CICIDS-2017

In addition, despite the fact that many studies use Information Gain as a feature selection approach, there are few talks about how to calculate the minimal weight either or rank score from the Information Gain result. This score indicates how important the qualities are to the class label. In studies [15, 16], researchers utilize a score characteristic of greater than 0.4 and greater than 0.001, as a result. Meanwhile, research [10] considers a weight score of 0.8 as a minimum. On the other hand, researchers [17 - 21] used a more formal procedure to locate the most accurate features by removing them one at a time. Such work takes a long time, especially if the dataset has a large number of features.

METHODOLOGY

The dataset-specification, experimentalism, feature selection method, classification algorithms, and experimental instruments are all covered in this part.

Table 2. Summary of the CICIDS-2017 dataset.

Attack Type	Record Count
Benign	2271320
DoS Hulk	230124
PortScan	158804
DdoS	128025
DoS GoldenEyes	10293
FTP-Patator	7935
SSH-Patator	5897
DoS Slowloris	5796
DoS Slowhttptest	5499
Bot	1956
Web Attack - Brute Force	1507
Web Attack - XSS	652
Infiltration	36
Web Attack Injection	21

(Table 2) cont.....

Attack Type	Record Count
Heartbleed	11
Total	2827876

Dataset Specification

MachineLearningCSV data from the CICIDS-2017 dataset from the ISCX Consortium was used in this work. MachineLearningCSV is made up of eight packet monitoring events, each of which is separated by a Comma Separated Value(CSV). This dataset has a normal packet defined as a Benign packet and a normal packet. Table **2** shows detailed information about this dataset's attack. This dataset contains 14 known attacks, other than a benign packet and a normal packet. In this research, based on their packet properties, we examine complex elements that indicate malicious activities on modern networks. For instance, elements like SubFlow FwdBytes and TotalLength FwdPackage are required to identify Infiltration and Bot attack types present in CICIDS-2017 but not in NSL-KDD. DoS Hulk, DoE GoldenEye, and Heartbleed attacks must be identified by the BwdPacket in order to identify DDoS. The Init WinFwd Bytes capability is crucial for detecting Web-Attack, SSH-Patator, and FTP-Patator attacks. To recognize normal packets, package length attributes, MinBwd Package Length and FwdAverage Package Length features are necessary. As seen in Table **2**, CICIDS-2017 features more complex methods of attacks. The rationale for using the CICIDS-2017 dataset is that it closely reflects the current real-world network packet in the tests.

Experimentalism

In general, the experimental setups presented in Fig. (**1**) have four steps, which can be explained as follows:

1. This experiment uses only twenty percent of the MachineLearningCSV data from the dataset. Because the dataset contains redundant features, the unnecessary ones must be removed. The process of relabeling is then carried out. The remaining twenty percent of the MachineLearningCSV data is divided into seventy percent training data and thirty percent testing data.
2. Information Gain is used to do feature engineering on the training data. Then, based on respective weights, relevant features are grouped.
3. After that, each feature subset or feature grouping is designated a classification, using Naïve bayes, J48 classifiers, Random-Tree (RT), Naïve-Bayes (NB), and Random Forest (RF). The TPR, FPR, precision, recall, accuracy, percentage of

incorrectly categorized items, and execution time are all factors taken into account. For this investigation, the 10-fold cross-validation is applied.

4. Finally, Compare and examine the TPR, FPR, Precision, and other metrics. Recall, accuracy, and the percentage of erroneously classified items are all factors to consider. Besides the time it takes for each classifier algorithm to run all, this analysis uses 10-fold cross-validation to carry out the learning and testing phases.

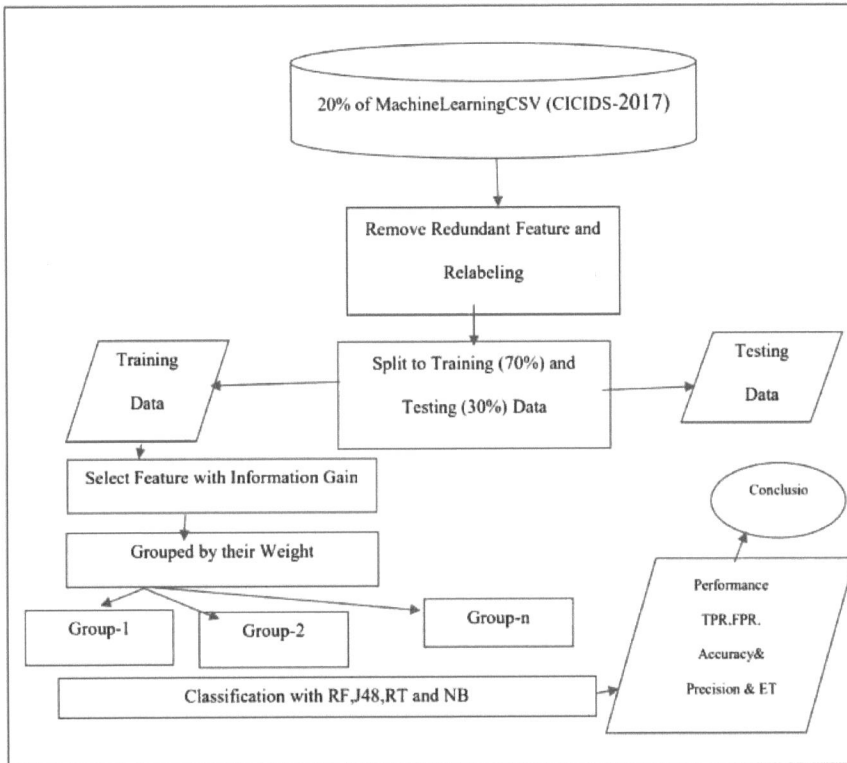

Fig. (1). Experimental Setup.

Information-Gain

The most common feature selection strategy is Information Gain. It is a feature selection based on the filter technique. Information Gain lowers noise by using a basic attribute rank and subsequently recognizes a feature that has been caused by irrelevant features. The majority of the data is organized by class. The most appealing aspect is determined by calculating the entropy of a feature. Entropy is a measure of the degree of disorder in a system. The measurement of entropy can be calculated in the way given below:

$$Entropy(S) = \sum_{i}^{c} -P_i \log_2 P_i \qquad (1)$$

c refers to the number of values in the classification class, and the Pi refers to the number of samples in the class. The Information Gain value is determined using the Entropy value as per equation (2).

$$Gain(S, A) = Entropy(S) - \sum_{values(A)} \frac{|S_V|}{|S|} Entropy(S_V) \qquad (2)$$

Where S stands for sample, A stands for attribute, v stands for a possible value for attribute A, and Values(A) stands for a set of possible values for A. The number of samples for value v is denoted by Sv. For all data samples, S is the number of samples, and Entropy (Sv) is the entropy for samples with a value of v.

For intruder detection, there are 77 features listed. The minimum weight was discovered by experimentation, and the Information Gain ranks the features according to their weight values. In this study, we propose that the attributes be ranked and grouped based on their minimum weight values. As a result, feature groups are formed, with each feature group having a distinct feature based on information gain, as illustrated in Table 6. Additionally, all feature groups will be examined using the classification techniques mentioned above. It allows us to determine which feature groups are efficient enough to be used for attack type classification.

Tools Used for Analysis

All simulations for this experiment are carried out on a computer running Windows 10 with an Intel Core i7 CPU clocked at 3.5 GHz and 12 GB of RAM. Analysis is done by using Rapid Miner, a machine learning program with a heap size of 3072 MB.

EXPERIMENTALISM

This part explains how to prepare the data for pre-processing, how to experiment with feature selection, classification and how the experiments worked out.

Table 3 depicts the distributed data labelled attack in the CICIDS 2017 dataset, where we used only twenty percent of the data. It shows the different instances for each attack among the 566239 total attacks.

Table 3. Distributed Data labelled attack in the CICIDS-2017 Dataset (20% of Machine LearningCSV data).

New Labels	Old Labels	# of Instances	Fraction to Majority Class	Fraction to Total Instance
Normal	Benign	454,396	100.00	80.25
Bot	Bot	367	0.081	0.06
Brute Force	FTP-Patator SSH-Patator	2,717	0.598	0.48
DoS/DdoS	DDoS, DoS, GoldenEye, DoS Hulk, DoS Slow, Httptest, DoS Slowloris, Heartbleed	76,445	16.82	13.50
Inflitration	Inflitration	6	0.001	0.00
PortScan	PortScan	31,882	7.061	5.63
Web Attack	Web Attack - Brute Force, Web Attack SQLInjection, Web Attack - XSS	426	0.094	0.08
Total Instances	-	566,239	-	-

Table **4** depicts the distributed data labelled attacks in the CICIDS 2017 dataset,

Table 4. Distributed Data from training and testing.

New Labels	Instances # of Training Data	Instances # of Testing Data
Normal	318,087	136,219
Bot	265	102
Brute Force	1,904	813
DoS/DdoS	53,427	23,018
Inflitration	5	1
PortScan	22,324	9,558
Web Attack	292	134
Total Instances	**396,304**	**169,845**

Data Preparation Process

Table **2** lists eight CSV files that are concatenated into one CSV file. This comma separated file needs to translate into the ARFF file in order to process the dataset with Rapid Miner software. Only 20% of the MachineLearningCSV data is used in the experiment. In this investigation, seventy-eight normal features and one class label were used.

The dataset has two "Fwd Header Length" features or columns that are redundant, so one of them must be eliminated. After deleting the duplicate features, there are only 77 features left to examine. According to the CICIDS-2017 data, data with a significant degree of class imbalance will have a low detection accuracy and a high rate of false alarms. By following their recommendations [2, 4, 10, 12-16, 18-21].

After relabeling the attack classes, the remaining 20% of MachineLearningCSV data is divided into two parts: 70% and 30%. The majority of the data is used for training, while the remainder is used for other purposes. As shown in Table **4**, 30% of the total is used for testing data, which made use of the 70:30 data segment.

It shows that using the 70:30 share of training and testing data yields the same degree of accuracy as using the 80:20 and 60:40 shares. The experimental outcome of using a 70:30 data component yielded great accuracy. As a result, the researchers divided the testing and training data 30:70 in this study. Despite the fact that the dataset has been modified to reflect a new attack classification, "infiltration" attacks have a very small portion of data in comparison to other types of attacks. After that, the data will be assessed using the feature selection approach.

Feature Selection Based On IG

The fundamental challenge in a big dataset, as indicated in Section 1, is dimensionality. By picking relevant characteristics, the feature selection technique decreases the dimensionality of data. The Information Gain calculates the entropies of the features to evaluate them. Weka software is used to implement feature selection in this work, and the procedure is depicted in algorithm 1. Table **5** shows the feature rank as a consequence of Information Gain's feature selection. The feature selection in this experiment is based on a filter based strategy, as indicated in sub-section experiment Results.

Table 5. Feature rank distributed by information gain.

No.	Feat.ID	Feature Names	Weight	No.	Feat.ID	Feature Names	Weight
1	41	Packet Length Std	0,638	40	17	Fwd Packet Length Std	0,280
2	13	Total Length of Bwd Packets	0,612	41	29	Bwd IAT Mean	0,271
3	65	Subflow Bwd Bytes	0,612	42	5	Fwd IAT Std	0,268
4	8	Destination Port	0,609	43	15	Fwd Packet Length Min	0,234
5	42	Packet Length Variance	0,577	44	38	Min Packet Length	0,231
6	20	Bwd Packet Length Mean	0,567	45	70	Active Mean	0,231

(Table 5) cont.....

7	54	Avg Bwd Segment Size	0,567	46	27	Fwd IAT Mean	0,229
8	18	Bwd Packet Length Max	0,560	47	73	Active Min	0,228
9	67	Init_Win_bytes_backward	0,554	48	69	Min_seg_size_forward	0,227
10	12	Total Length of Fwd Packets	0,546	49	72	Active Max	0,226
11	63	Subflow Fwd Bytes	0,546	50	31	Bwd IAT Min	0,226
12	66	Init_Win_bytes_forward	0,542	51	23	Flow IAT Min	0,216
13	52	Average Packet Size	0,535	52	76	Idle Max	0,205
14	40	Packet Length Mean	0,526	53	74	Idle Mean	0,197
15	39	Max Packet Length	0,512	54	77	Idle Min	0,195
16	14	Fwd Packet Length Max	0,495	55	68	Act_data_pkt_fwd	0,186
17	22	Flow IAT Max	0,467	56	6	Bwd IAT Std	0,179
18	36	Bwd Header Length	0,448	57	46	PSH Flag Count	0,106
19	9	Flow Duration	0,443	58	51	Down/Up Ratio	0,088
20	26	Fwd IAT Max	0,438	59	47	ACK Flag Count	0,069
21	55	Fwd Header Length	0,431	60	75	Idle Std	0,036
22	24	Fwd IAT Total	0,415	61	43	FIN Flag Count	0,033
23	25	Fwd IAT Mean	0,390	62	48	URG Flag Count	0,028
24	21	Flow IAT Mean	0,379	63	71	Active Std	0,025
25	2	Flow Byte/s	0,379	64	44	SYN Flag Count	0,012
26	1	Bwd Packet Length Std	0,360	65	32	Fwd PSH Flags	0,012
27	64	Subflow Bwd Packets	0,355	66	45	RST Flag Count	0
28	11	Total Backward Packets	0,355	67	50	ECE Flag Count	0
29	16	Fwd Packet Length Mean	0,351	68	61	Bwd Avg Bulk Rate	0
30	53	Avg Fwd Segment Size	0,351	69	49	CWE Flag Count	0
31	19	Bwd Packet Length Min	0,324	70	57	Fwd Avg Packet/Bulk	0
32	3	Flow Packets/s	0,311	71	56	Fwd Avg Bytes/Bulk	0
33	37	Fwd Packets/s	0,309	72	34	Fwd URG Flags	0
34	30	Bwd IAT Mean	0,306	73	33	Bwd PSH Flags	0
35	7	Bwd Packets/s	0,306	74	35	Bwd URG Flags	0
36	10	Total Fwd Packets/s	0,291	75	60	Bwd Avg Packets/Bulk	0
37	62	Subflow Fwd Packets	0,291	76	58	Fwd Avg Bulk Rate	0
38	28	Bwd IAT Total	0,287	77	59	Bwd Avg Bytes/Bulk	0
39	4	Flow IAT Std	-	-	-	-	-

Where we found labeled attacks with the featured ID.

In other words, feature selection occurs throughout the weight scores, with features grouped according to their weight score. There are seven groups of features presented in Table **6**, which we refer to as new features of subsets.

Table 6. Feature selected based on information gain.

Feature Weight	Number of Selected Feature	Selected Features (New Feature Subset)
>0.6	4	41, 13, 65, 8
>0.5	15	41, 13, 65, 8, 42, 20, 54, 18, 67, 12, 63, 66, 52, 40, 39
>0.4	22	41, 13, 65, 8, 42, 20, 54, 18, 67, 12, 63, 66, 52, 40, 39, 14, 22, 36, 9, 26, 55, 24
>0.3	35	41, 13, 65, 8, 42, 20, 54, 18, 67, 12, 63, 66, 52, 40, 39, 14, 22, 36, 9, 26, 55, 24, 25, 21, 2 , 1, 64, 11, 16, 53, 19, 3, 37, 30, 7
>0.2	52	41, 13, 65, 8, 42, 20, 54, 18, 67, 12, 63, 66, 52, 40, 39, 14, 22, 36, 9, 26, 55, 24, 25, 21, 2 , 1, 64, 11, 16, 53, 19, 3, 37, 30, 7, 10, 62, 28, 4, 17, 29, 5, 15, 38, 70, 27, 73, 69, 72, 31, 23, 76
>0.1	57	41, 13, 65, 8, 42, 20, 54, 18, 67, 12, 63, 66, 52, 40, 39, 14, 22, 36, 9, 26, 55, 24, 25, 21, 2 , 1, 64, 11, 16, 53, 19, 3, 37, 30, 7, 10, 62, 28, 4, 17, 29, 5, 15, 38, 70, 27, 73, 69, 72, 31, 23, 76, 74, 77, 68, 6, 46
All	77	*All Feature*

Experimental Result

The performance of the feature selection provided by Information Gain and the five classification algorithms is evaluated using seven measurement metrics: True Positive Rate (TPR), False Positive Rate (FPR), Precision, Recall, Accuracy, percentage of incorrectly classified, and execution time. Throughout the training period, the execution time is measured (the time measured from the classification process starts until the classification process stops). Each feature subset in the experiment is classified by the RT, BN, RT, NB, and J48 classifiers.

The below given algorithm depicts the overall process. This research uses 10-fold cross-validation to assess the performance of classification algorithms. We used 10-fold cross-validation, which saves the time maintaining the classification's performance.

In terms of precision, algorithms. As a result, the input dataset will be randomly divided into ten folds with the same dimension's size. Cross-validation is used for each of the ten fold data. There are nine folds for training and one fold for testing. This is how it works, repeat for a total of ten times until each fold is a test fold.

Algorithm
1: **procedure** Start_Process()
2: **Input:** Fr D Feature_Ranked_data
3: **Output:** Features-Subsets, TPR, FPR, Accuracy, Recall, Precision
4: **dimensionality** reduction 77 features to n features based on a feature weight
5: **For** every feature Fr in Feature_Ranked-data
6: **Start to Select feature** with Feature Weight and store on Feature-Groups
7: Group1 D all feature with weight >D 0:7
8: Group2 D all feature with weight >D 0:6
9: Group3 D all feature with weight >D 0:5
10: Group4 D all feature with weight >D 0:4
11: Group5 D all feature with weight >D 0:3
12: Group6 D all feature with weight >D 0:2
13: Group7 D all features
14: **For each** Feature-groups
15: **Feed** Selected Features to RF, BN, RT, NB, J48 using CICIDS-2017-20%
16: **Apply Classifier**
10: C1 D Random Forest model accuracy
11: C2 D Bayes Network model accuracy
12: C3 D Random Tree model accuracy
13: C4 D Naïve Bayes model accuracy
14: C5 D J48 model accuracy
15: **Calculate** TPR, FPR Accuracy, Recall, Precision
16: **Compare** the Accuracy of C1, C2, C3, C4 and C5

Table 7 shows the results of classifiers utilizing four (4) features chosen by Information Gain. When compared to other classifiers, the RF and RT had the highest accuracy of 96.48 percent. RF, on the other hand, has a NaN value. The term "NaN" stands for "Not a Number" or "Undefined." In comparison to other classifiers, NB can detect DoS/DDoS attacks with a TPR of 0.999, but it has a low TPR for detecting Normal and Infiltration packets. Surprisingly, BN has the lowest FPR of the group, at 0.010. Overall, the classifiers can only detect DoS/DDoS, PortScan, and Brute Force assaults using only four (4) specified features. Only NB suffers from this in the case of the normal packet.

Table 7. Performance metrics using four features.

Detection	RF	BN	RT	NB	J48
Normal	0.960	0.943	0.960	0.174	0.961
DoS/DDoS	0.992	0.996	0.992	0.999	0.991
Port Scan	0.995	0.992	0.995	0.983	0.995
Bot	0.438	0.642	0.430	0.687	0.381
Web Attack	0.072	0.031	0.072	0.000	0.072
Infiltration	0.000	0.000	0.400	0.400	0.000
Brute Force	0.792	0.991	0.792	1.000	0.790
Recall	0.965	0.962	0.970	0.903	NaN
Precision	NaN	0.953	0.965	0.335	0.965
FPR	0.016	0.010	0.016	0.026	0.016

Table **8** summarizes the results of classifiers with 15 features. When compared to other classifiers, the RF has the highest accuracy of 99.81 percent. The results demonstrate that RF, RT, and J48 are capable of recognizing normal, DoS/DdoS, Bot, and Brute Force packets, however, they are less capable of detecting Web Attack and Infiltration packets. Furthermore, RF, RT, and J48 have low FPRs of 0.005, with BN having the lowest FPR of 0.002. With a rating of 0.998, the RF, RT, and J48 have good Precision and Recall. The performance of the classifiers with 22 selected features is then given in Table **9**. In comparison to the others, the result shows that RF has the highest accuracy of 99.86 percent. This classifier, too, has a high recall of 0.999, and unfortunately, the accuracy result suggests a NaN due to the low FPR value of 0.003. RF, on the other hand, is unable to detect infiltration using the 22 features specified; nonetheless, all classifiers have a high TPR for detecting DoS/DDoS. PortScan and Brute Force are two programs that can be used to scan a network for vulnerabilities. Only NB has a low TPR for normal packet RF. BN, RT, and J48 achieve good TPR for normal packet RF. Table **10** shows the results of the classifiers using 35 different features. RF has the highest accuracy of 99.83 percent, the highest recall of 0.998, and the lowest FPR of 0.004. Regardless, the precision is recorded as NaN. This result demonstrates that RF is unable to detect infiltration. Surprisingly, NB outperforms other approaches with a performance of 70.84 percent accuracy, despite the fact that this is lower than other ways. It does, however, have good precision with a value of 0.923. The result of 52 features is mentioned in Table **11**.

Table 8. Measures of performance with 15 features.

Detection	RF	BN	RT	NB	J48
Normal	0.999	0.874	0.999	0.304	0.999
DoS/DDoS	0.999	0.97	0.999	0.965	0.999
Port Scan	0.997	0.995	0.997	0.992	0.997
Bot	0.706	0.985	0.725	0.457	0.713
Web Attack	0.116	0.993	0.116	0.829	0.110
Infiltration	0.200	0.400	0.600	0.600	0
Brute Force	0.995	0.996	0.995	0.999	0.996
Recall	0.998	0.996	0.998	0.436	0.998
Precision	0.998	0.895	0.998	0.913	0.998
FPR	0.005	0.002	0.005	0.031	0.005

Table 9. Measures of performance with 22 features.

Detection	RF	BN	RT	NB	J48
Normal	0.999	0.927	0.999	0.358	0.999
DoS/DDoS	0.999	0.981	0.997	0.723	0.999
Port Scan	0.996	0.992	0.994	0.991	0.999
Bot	0.762	0.989	0.777	0.570	0.698
Web Attack	0.788	0.986	0.743	0.846	0.130
Infiltration	0	0.600	0.400	0.800	0
Brute Force	0.997	0.994	0.996	0.983	0.995
Recall	0.999	0.938	0.998	0.447	0.998
Precision	NaN	0.995	0.998	0.925	NaN
FPR	0.003	0.004	0.004	0.017	0.004

Table 10. Measures of performance with 35 features.

Detection	RF	BN	RT	NB	J48
Normal	0.999	0.92	0.998	0.653	0.999
DoS/DDoS	0.998	0.983	0.998	0.673	0.999
Port Scan	0.994	0.991	0.993	0.989	0.999
Bot	0.713	0.985	0.755	0.494	0.691
Web Attack	0.651	0.990	0.716	0.955	0.116
Infiltration	0.000	0.600	0.200	0.800	0.200
Brute Force	0.993	0.989	0.993	0.947	0.993

(Table 10) cont.....

Recall	0.998	0.933	0.998	0.708	0.998
Precision	NaN	0.993	0.998	0.923	0.998
FPR	0.004	0.006	0.004	0.013	0.004

Table 11. Measures of performance with 52 features.

Detection	RF	BN	RT	NB	J48
Normal	0.999	0.932	0.998	0.400	0.999
DoS/DDoS	0.998	0.978	0.997	0.715	0.999
Port Scan	0.994	0.991	0.993	0.931	0.999
Bot	0.668	0.989	0.732	0.774	0.698
Web Attack	0.942	0.990	0.925	0.993	0.949
Infiltration	0.000	1.000	0.000	0.800	0.000
Brute Force	0.993	0.994	0.992	0.963	0.993
Recall	0.998	0.942	0.998	0.476	0.999
Precision	NaN	0.994	0.998	0.880	0.999
FPR	0.004	0.009	0.004	0.035	0.002

When compared to other classifiers, J48 has a better performance with an accuracy of 99.87 percent, a recall of 0.999, a precision of 0.999, and a low FPR of 0.002. Table **12** shows the results of classifiers utilizing 57 different features. With good TPR values, BN can detect all types of packets.

Table 12. Measures of performance with 57 features.

Detection	RF	BN	RT	NB	J48
Normal	0.999	0.932	0.999	0.358	0.999
DoS/DDoS	0.998	0.973	0.997	0.724	0.999
Port Scan	0.994	0.991	0.993	0.489	0.999
Bot	0.668	0.989	0.751	0.777	0.721
Web Attack	0.932	0.990	0.911	0.993	0.949
Infiltration	0.000	1.000	0.200	0.800	0.000
Brute Force	0.993	0.994	0.990	0.963	0.993
Recall	0.998	0.942	0.998	0.871	0.999
Precision	NaN	0.994	0.998	0.871	0.999
FPR	0.004	0.011	0.004	0.037	0.002

Table **11** shows 52-feature performance metric. Table **12** presents 57-feature performance metric. Table **13** shows measures of performance for all 77 features. Finally, Table **13** summarizes the results of classifiers that use all features. BN can detect all types of packets with a high TPR by utilizing all features. Tables **11**, **12**, and **13** show that RF, RT, and J48 with 53, 57, and all features are capable of detecting normal, Dos/DDoS, Brute Force, and Bot assaults packets. RF, RT, and J48, on the other hand, have difficulty detecting Infiltration attack packets, although BN and NB have a high ability to do so.

Table 13. Measures of performance for all 77 features.

Detection	RF	BN	RT	NB	J48
Normal	0.999	0.940	0.998	0.333	0.999
DoS/DDoS	0.998	0.974	0.996	0.731	0.999
Port Scan	0.994	0.991	0.993	0.660	0.999
Bot	0.653	0.989	0.675	0.774	0.740
Web Attack	0.935	0.990	0.894	0.983	0.966
Infiltration	0.000	1.000	0.200	0.800	0.000
Brute Force	0.994	0.995	0.992	0.979	0.995
Recall	0.998	0.948	0.997	0.409	0.999
Precision	NaN	0.993	0.997	0.874	NaN
FPR	0.004	0.010	0.005	0.040	0.002

Experimental Analysis

In the experiments, the suggested Information Gain feature selection produces a ranking for the features based on their weight values. Higher weighted features signify more relevant and significant aspects of an attack. As can be seen in Table **5**, the experiment yielded the top four traits (out of 77) and their scores. As a result, features with IDs 41, 13, 65, and 8 are the most relevant and significant for detecting any attacks and can be found in any of the feature subsets. Finally, the RF, J48, BN, and RT classifiers can recognize normal packet, Brute Force, DoS/DDoS, Port Scan and Web attacks packet utilizing the 77,52, and 35 feature subsets. The fact that the classifiers use a strong decision tree learning technique confirms this conclusion.

In the example of detecting an Infiltration attack packet, NB can detect with a TPR of 0.8 using features subsets 22 and 35, and perfectly detect (with a TPR of 1.0) using features subsets 52, 57, and 77. The reason for this is that significant features suggesting an Infiltration attack packet exist in the 52, 55, and 77 feature subsets. The little amount of this type of attack packet in the dataset may result in poor detection performance. As noted in item 4.A, CCIDS-2017 has skewed data, which makes spotting anomalies/attacks difficult. The Infiltration attack packet cannot be detected by other classifiers like BN, RF, RT, and J48.

Using the characteristics subset of 4, all classifiers are unable to recognize the Web Attack packet, the same like they are unable to detect the Infiltration attack. Then only BN and NB classifiers will be able to get involved and detect a web attack packet using a subset of 15 features 0.993 and 0.829, respectively, are the TPR values. RF, BN, RT, and J48 are the bot attack packet detection algorithms are able to detect data using subsets of features, However, the TPR values are lower.

The findings of the experiment reveal that the type and number of selected features have a significant impact on the detection performance. Fig. (**2**) depicts a summary of the classifiers' findings. The amount of selected features as a result of the designed Learning Information Gain has an impact on accuracy. The RF and RT, and the proposed Information Gain obtain the greatest accuracy of 99.86 percent in RF and 99.78 percent for RT, based on a 22-feature subset. The proposed Information Gain, but from the other hand with 35 picked, enhances NB's accuracy by up to 70.84 percent features. There is no significant difference between BN and J48, when compared to using all of the features in the analysis. Apart from consistency, several aspects have an effect on the FPR, as illustrated in Fig. (**3**). In terms of the FPR, 22 selected features had a 0.003 effect on RF's FPR. When compared to the utilization of all features, it is slightly lower. In the case of BN, 15 selected features had a 0.002 effect on FPR. Among the number of selected features, this has the lowest FPR. In the same way as 22 selected features affected RF's FPR by up to 0.004, 22 selected features affected RT's FPR by up to 0.004. The proposed feature selection for Information Gain has a significant impact on NB's FPR. Fourteen, twenty-two, and thirty-five feature subsets have an impact on this impact. When compared to all features subset, the proposed Information Gain does not diminish FPR for J48 but rather boosts it. The effect of execution duration on the selected feature process is also investigated in this study. Fig. (**4**) summarizes the time it took to obtain each feature subset using RF, J48, RT, BN and NB. The relevant selected features procedure has a significant impact. RT and NB both have a very short execution time. In general, a number of features to assess more time are necessary for the execution.

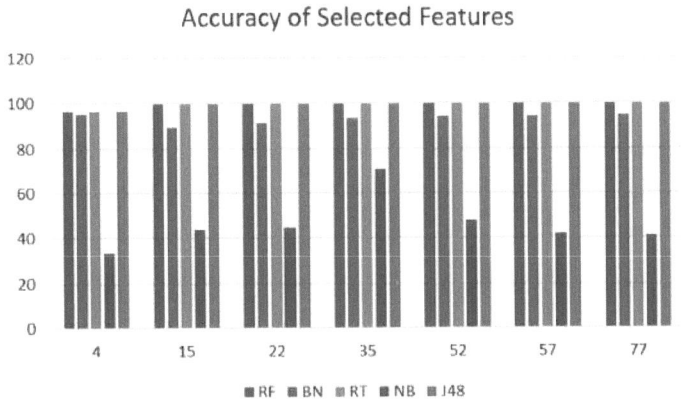

Fig. (2). Exactness of chosen features.

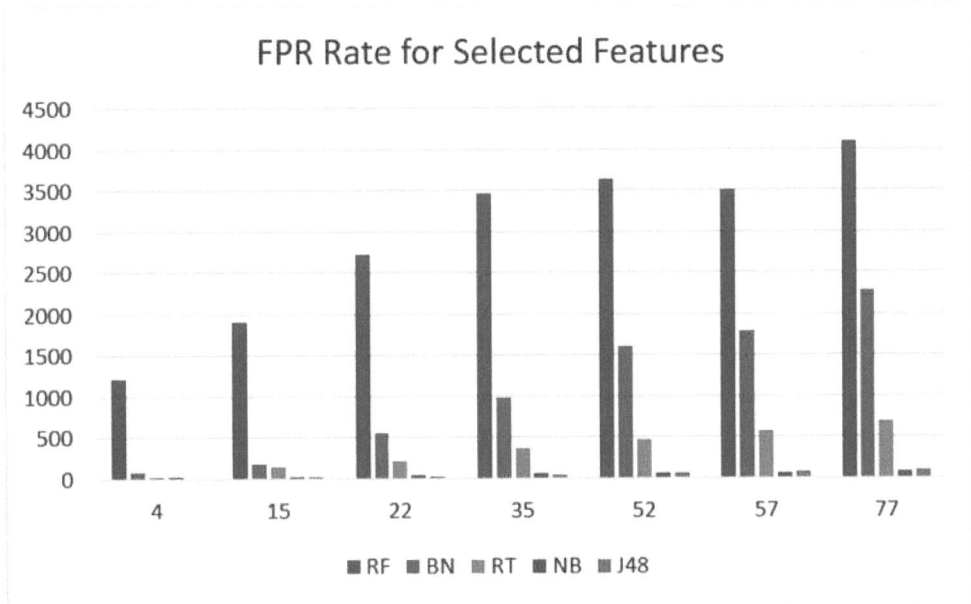

Fig. (3). FPR Rate Selected Features.

Execution Time

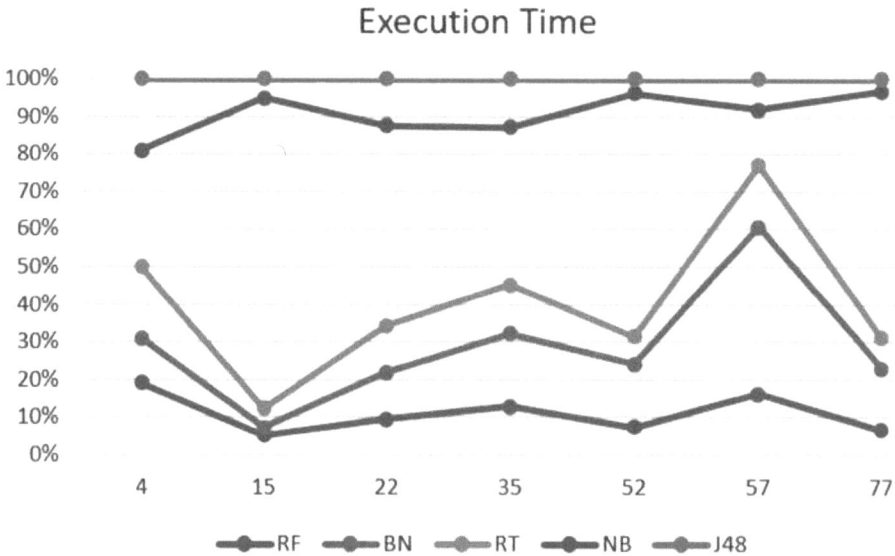

Fig. (4). Execution Time for Selected Features.

CONCLUSION

This paper presents the results of a study that looked at the impact of feature selection on improving abnormality detection. Its name comes from its capacity to determine the weight of features. When applying feature subsets of 35, 22 and 15, the Information Gain given with RF classier outperforms the contest. J48, on the other hand, succeeds when feature subsets 77, 52, and 57 are used. The experiment shows that BN can detect all packets using 52, 57, and 77 feature subsets, despite having worse accuracy than RF and J48. Additionally, the research' findings demonstrate that the selected features lower the FPR level, particularly for BN. Based on their weight values, the proposed Information Gain creates ranked features. Based on their weight values, the proposed Information Gain generates ranked features. Defining the minimal weight value is still required. The authors intend to evaluate a range of feature selection techniques in order to identify the best feature selection mechanism. In the future, each feature subset that affects each kind of attack will be investigated.

REFERENCES

[1]   R. Panigrahi, and S. Borah, "A detailed analysis of CICIDS2017 dataset for designing Intrusion Detection Systems", *Int. J. Eng. Technol,* vol. 7, no. 24, pp. 479-482, 2018.
[http://dx.doi.org/10.14419/ijet.v7i3.24.22797]

[2]   P. Tao, Z. Sun, and Z. Sun, "An Improved Intrusion Detection Algorithm Based on GA and SVM", *IEEE Access,* vol. 6, pp. 13624-13631, 2018.
[http://dx.doi.org/10.1109/ACCESS.2018.2810198]

[3]   R.K. Singh, S. Dalal, V.K. Chauhan, and D. Kumar, "Optimization of FAR in Intrusion Detection System by Using Random Forest Algorithm", *SSRN,* pp. 3-6, 2019.
[http://dx.doi.org/10.2139/ssrn.3350276]

[4]   T.A. Alhaj, M.M. Siraj, A. Zainal, H.T. Elshoush, and F. Elhaj, "Feature selection using information gain for improved structural-based alert correlation", *PLoS One,* vol. 11, no. 11, p. e0166017, 2016.
[http://dx.doi.org/10.1371/journal.pone.0166017] [PMID: 27893821]

[5]   A. Niranjan, D.H. Nutan, A. Nitish, P.D. Shenoy, and K.R. Venugopal, "ERCR TV: Ensemble of random committee and random tree for efficient anomaly classification using voting", In: *3rd International Conference for Convergence in Technology (I2CT),* 2018.
[http://dx.doi.org/10.1109/I2CT.2018.8529797]

[6]   A.S. Eesa, Z. Orman, and A.M.A. Brifcani, "A novel feature-selection approach based on the cuttlefish optimization algorithm for intrusion detection systems", *Expert Syst. Appl.,* vol. 42, no. 5, pp. 2670-2679, 2015.
[http://dx.doi.org/10.1016/j.eswa.2014.11.009]

[7]   K. Rai, M.S. Devi, and A. Guleria, "Decision tree based algorithm for intrusion detection", *Int. J. Adv. Netw. Appl.,* vol. 7, no. 4, pp. 2828-2834, 2016.

[8]   I. Sharafaldin, A.H. Lashkari, and A.A. Ghorbani, "Toward generating a new intrusion detection dataset and intrusion packet characterization", *4th International Conference on Information Systems Security and Privacy,* 2018.
[http://dx.doi.org/10.5220/0006639801080116]

[9]   A.L. Buczak, and E. Guven, "A Survey of Data Mining and Machine Learning Methods for Cyber Security Intrusion Detection", *IEEE Commun. Surv. Tutor.,* vol. 18, no. 2, pp. 1153-1176, 2016.
[http://dx.doi.org/10.1109/COMST.2015.2494502]

[10]   M.A. Ambusaidi, X. He, P. Nanda, and Z. Tan, "Building an intrusion detection system using a filter-based feature selection algorithm", *IEEE Trans. Comput.,* vol. 65, no. 10, pp. 2986-2998, 2016.
[http://dx.doi.org/10.1109/TC.2016.2519914]

[11]   M. Reazul, A. Rahman, and T. Samad, "A Network Intrusion Detection Framework based on Bayesian Network using Wrapper Approach", *Int. J. Comput. Appl.,* vol. 166, no. 4, pp. 13-17, 2017.
[http://dx.doi.org/10.5120/ijca2017913992]

[12]   F. Chen, Z. Ye, C. Wang, L. Yan, and R. Wang, "A feature selection approach for network intrusion detection based on tree-seed algorithm and k-nearest neighbor", *Proc. 2018 IEEE 4th Int. Symp. Wirel. Syst. within Int. Conf. Intell. Data Acquis. Adv. Comput. Syst. IDAACS-SWS 2018,* pp. 68-72, 2018.
[http://dx.doi.org/10.1109/IDAACS-SWS.2018.8525522]

[13]   D. Kurniabudi, D. Stiawan, Darmawijoyo, M.Y. Bin Idris, A.M. Bamhdi, and R. Budiarto, "CICIDS-2017 Dataset Feature Analysis With Information Gain for Anomaly Detection", *IEEE Access,* vol. 8, pp. 132911-132921, 2020.
[http://dx.doi.org/10.1109/ACCESS.2020.3009843]

[14]   H. Peng, C. Ying, S. Tan, B. Hu, and Z. Sun, "An Improved Feature Selection Algorithm Based on Ant Colony Optimization", *IEEE Access,* vol. 6, pp. 69203-69209, 2018.
[http://dx.doi.org/10.1109/ACCESS.2018.2879583]

[15]   A.A.A. Hadi, and A-A. Al-Furat, "Performance analysis of big data intrusion detection system over random Forest algorithm", *Int. J. Appl. Eng. Res.,* vol. 13, no. 2, pp. 1520-1527, 2018.

[16]   A. Yulianto, P. Sukarno, and N.A. Suwastika, "Improving AdaBoost-based Intrusion Detection System (IDS) Performance on CIC IDS 2017 Dataset", *J. Phys. Conf. Ser.,* vol. 1192, no. 1, p. 012018, 2019.

[http://dx.doi.org/10.1088/1742-6596/1192/1/012018]

[17]   S. Aljawarneh, M.B. Yassein, and M. Aljundi, "An enhanced J48 classification algorithm for the anomaly intrusion detection systems", *Cluster Comput.,* vol. 22, no. S5, pp. 10549-10565, 2019. [http://dx.doi.org/10.1007/s10586-017-1109-8]

[18]   S. Aljawarneh, M. Aldwairi, and M.B. Yassein, "Anomaly-based intrusion detection system through feature selection analysis and building hybrid efficient model", *J. Comput. Sci.,* vol. 25, pp. 152-160, 2018. [http://dx.doi.org/10.1016/j.jocs.2017.03.006]

[19]   I.M. Akashdeep, I. Manzoor, and N. Kumar, "A feature reduced intrusion detection system using ANN classifier", *Expert Syst. Appl.,* vol. 88, pp. 249-257, 2017. [http://dx.doi.org/10.1016/j.eswa.2017.07.005]

[20]   M. Abualkibash, "Machine Learning in Network Security Using KNIME Analytics", *International Journal of Network Security & Its Applications,* vol. 11, no. 5, pp. 1-14, 2019. [http://dx.doi.org/10.5121/ijnsa.2019.11501]

[21]   "Advanced computing, networking and informatics- volume 1", In: *Smart Innov. Syst. Technol.* vol. 27. Springer, 2014, pp. 205-206. [http://dx.doi.org/10.1007/978-3-319-07353-8]

CHAPTER 11

Machine Learning Based Crop Recommendation System

Keerti Adapa[1] and **Sudheer Hanumanthakari**[1,*]

[1] *Faculty of Science and Technology, ICFAI Foundation for Higher Education, Hyderabad, India*

Abstract: Agriculture is very important in the Indian economy. Nowadays, due to the change in climate and the increase in global warming, the weather is an unpredictable variable. So, the most common issue that Indian farmers encounter is that they fail to identify the best-suited and appropriate crop for their soil using conventional methods. As a result, they experience a significant drop in production. This is a big problem in a country where farming employs over 58 percent of the population and results in low crop production. To overcome this issue, a model is built using machine learning which has a better system to guide the farmers, and it is a modern agricultural strategy for selecting the best crop by considering all the factors like nitrogen, phosphorus, potassium percentages, temperature, humidity, rainfall, and ph value. This paper proposes the use of machine learning techniques such as logistic regression, decision tree, KNN (k-Nearest Neighbours) and Naive Bayes to determine the best-suited crop based on attributes of soil and environmental factors. In the end, an accuracy of 96.36 percent from the logistic regression, 99.54 percent from the decision tree, 98.03 percent from the k-nearest neighbours and 99.09 percent from the naive Bayes is obtained, resulting in the decision tree having the highest accuracy with 99.54 percent. This paper gives an extensive Exploratory Data Analysis (EDA) on the Crop recommendation Dataset and builds an appropriate Machine Learning Model that will help farmers predict their suitable crops based on their parameters.

Keywords: Accuracy, Crop recommendation, Dataset, Data pre-processing, Decision tree, Humidity, k-nearest neighbours (KNN) algorithm, Logistic Regression, Machine learning (ML), ML algorithms, Naive Bayes algorithm, Ph value, Python, Rainfall, SciKit-learn, Soil NPK percentages, Streamlit, Temperature.

INTRODUCTION

Agriculture is a standout amongst the most vital divisions in numerous countries. Maintaining sustainable agriculture in a country like India, where agriculture and

---

* **Corresponding author Sudheer Hanumanthakari:** Faculty of Science and Technology, ICFAI Foundation for Higher Education, Hyderabad, India; E-mail: hsudheer@ifheindia,org

Hemachandran K., Raul V. Rodriguez, Umashankar Subramaniam & Valentina Emilia Balas (Eds.)

allied sectors account for roughly 20% of the country's Gross Value Added (GVA), is important to meet rising demand [1]. In general, a farmer's selection on which crop to produce is impacted by his intuition as well as other non-essential variables such as making quick money, being oblivious of market demand, overestimating a soil's potential to support a specific crop, and so on [1]. A farmer's erroneous decision could put a substantial burden on his family's financial situation, and such an error would have bad consequences not only for the farmer's family but for the entire economy of an area. Despite the fact that many initiatives have been taken to reduce crop loss, traditional methods have their own set of disadvantages. As a result, rather than practising traditional farming methods, it is critical to adopt modern agricultural methods that make use of technology. Hence there is a need to develop a user-friendly scientific procedure for farmers which helps them to select the crop based on scientific attributes line, type of soil, temperature and rainfall forecast, nitrogen in soil, pH values, *etc* [2].

Effect of Soil Types on Crop Production

Soil is made up of inorganic particles and organic stuff, and it offers structural support to agricultural plants. Soil is classified based on its chemical and physical properties. Leaching, weathering, and microbiological activity all work together to create a diverse spectrum of soil types. For agricultural productivity, each variety has distinct strengths and limitations. The texture of the soil is determined by the combination of mineral fractions (gravel, sand, silt, and clay particles) and organic matter fractions. Few crops, such as rice cultivation, require water wetness; hence clay soil is preferred over sandy soil.

Soil colour can illustrate the organic nutrient, parent material, the degree of weathering, and the features of the soil drainage. The key indicator of how soils drain is the colour of the soil. White sands, for example, have a lighter colour that indicates low fertility. Darker soils (such as black clays) are, on the other hand, highly fertile. The percentage of nitrogen, phosphorous and potassium present in the soil is the major factor that affects crop production. Nitrogen helps in the growth of leafs, phosphorus helps in the development of new roots and seed growth, and potassium aids in the formation of strong stems and the rapid growth of plants.

Effect of Rainfall on Crop Production

Rainfall has a significant impact on soil. It may also predict how quickly a crop will grow from seed to maturity, as well as when it will be ready to harvest. A healthy rain-water balance combined with efficient irrigation can result in faster-growing plants and the time between seeding and harvest. Excessive rainfall can have a variety of effects on crop yield. It leads to physical damage to the crop, and root growth is restrained. It causes nutrient loss and oxygen insufficiency, resulting in poor growth and overall health. Overwatering or too much rain can also encourage the growth of bacteria, fungi, and mould in the soil [3].

Role of Temperature in Crop Production

Plant growth and development are also influenced by temperature. The plant is affected by temperature both in the short and long term. The ideal temperature for a plant is determined by a number of factors. Plant growth is also influenced by its stage of development and how it reacts to temperature changes over time till harvesting [4].

Crop Recommendation System

To eliminate the aforementioned drawbacks, an intelligent crop recommendation system is proposed; it takes into account all relevant parameters, such as type and condition of the soil, rainfall, temperature, and geographical location, to anticipate crop suitability. This framework is generally worried with filling the essential role of Agro-Consultant, giving yield suggestions to farmer's algorithms. In general, recommendation systems can be characterized as a class of programming which assists clients with getting the most reasonable product as indicated by their inclinations, needs, or tastes. It applies information disclosure methods to the issue of customized suggestions for data, products or services generally known as e-commerce, social media and content-based sites [5, 6].

The use of recommendation system machine learning models will help in the farming sector. By picking the right factors, the most valuable information can be developed using the Machine Learning (ML) models, and this produced information helps the farmers by recommending the right harvest to be planted. The fundamental point of the Crop Recommendation model is to propose the exact crop for a specific field region. By picking reasonable harvests for the field region, we can limit the deficiency of yields. Appropriate algorithms with specific highlights must be picked since the level of accuracy in crop recommendation varies depending on the algorithm used. Machine Learning is tracked down to be the best innovation for anticipating appropriate crops and their yield. The ML models are developed using Python programming language because it is widely acknowledged in the Machine Learning field [7].

THEORETICAL BACKGROUND

Overview of Machine Learning

Machine learning is a part of artificial intelligence in which the system learns from examples. In machine learning, the learning process begins with data, examples or instructions, so that we can seek patterns in data and make better decisions based on the examples we provide in the future. The primary goal of machine learning is to enable computers to learn on their own and adapt their actions to improve the program's accuracy and usefulness without the need for human interaction. The traditional definition of writing computer programmes is automating the procedures that must be done on input data in order to produce output artefacts. Almost always, they are linear, procedural and logical.

SciKit-learn

Scikit-learn is an open-source system mastering library constructed for Python. Considering its launch in 2007, scikit-study has become one of the most famous open-source machine learning libraries. Scikit-research provides algorithms for plenty of machines gaining knowledge of responsibilities along with category, regression, dimensionality discount and clustering. Scikit-learnis constructed on mature Python libraries along with NumPy, SciPy, and matplotlib. In this paper, we used the Scikit-learn library for the implementation of Machine learning models due to the following reasons: We already have got some familiarity and exposure to Python and, for this reason, have a smaller getting to know curve. Both Python and scikit-learn have extraordinary documentation and tutorials available online. A variety of classic systems gaining knowledge of algorithms come with scikit-learn, and the steady styles use the exclusive models, *i.e.,* every version can be used with the same basic commands for putting in the facts, schooling the model and using that version for prediction. This makes it simpler to try a number of machine learning algorithms on equal records.

Dataset

For crop recommendation, the dataset downloaded Named "Crop_recom-mendation.csv", from Kaggle is used, which has 2200 records and was created by augmenting existing data of rainfall, climate, and fertilizers available in India. The attributes considered were nitrogen, phosphorous, potassium, temperature, humidity, pH, and rainfall. The aforementioned parameters are critical for the growth of crops to reach their maximum potential.

Soil is the anchor of the roots. The level of acidity or alkalinity (pH) is one of the important attributes which determines the soil nutrients. The pH of soil can influence the activity of soil microorganisms as well as the level of exchangeable aluminium. These essential nutrients, nitrogen, phosphorus, and potassium, play an important role in plant nutrition. Nitrogen is required for plants to grow healthy and to be nutritious to consume once harvested. Phosphorus is linked to the plant's ability to use and store energy, including photosynthesis. Potassium helps plants to resist diseases and is essential for increasing crop yield and overall quality. Potassium also protects the plant by strengthening its root system and preventing wilt when the weather is cold or dry. Rainfall is also an important factor in predicting yield. Because each harvest has a unique water requirement, we took this into account while predicting the crop. Temperature is the most important factor influencing plant improvement charge.

Data Pre-processing

Data pre-processing is an important step in Machine Learning as it helps in improving the data quality and promotes the extraction of meaningful insights from data. Since machine learning models are entirely based on mathematics and numbers, the presence of a categorical variable in our dataset may cause issues while building the model. As a result, these categorical variables must be numerically encoded. To accomplish this, we will employ the LabelEncoder() class from the pre-processing library. Our dataset is divided into two parts: training and testing sets. This is a crucial step in data pre-processing because it enables us to improve the performance of our machine-learning model. A programmer would always strive to build a machine-learning model that works well with both training and test datasets. The training data accounts for approximately 70% of the dataset, while the test data accounts for approximately 30% of the dataset.

Streamlit

Streamlit is a Python framework that allows you to create and share web apps for machine learning projects. With just a few lines of code, you can create and deploy your data science solution in minutes. Several open-source projects that have used streamlit for analytics and machine learning can be found in the streamlit gallery. Streamlit apps are entirely served over HTTPS. 256-bit encryption is used to encrypt all data sent to or from streamlit over the public internet. Because of the layout features that were introduced relatively recently, Streamlit has become a viable general-purpose tool for creating web pages. The best thing about streamlit is that it does not necessitate any prior knowledge of web development. You're good to go if you know python

MACHINE LEARNING ALGORITHMS

Machine Learning algorithms used in the recommendation system are:

Logistic Regression

Logistic Regression method is used to perform regression analysis used for prediction of the categorical dependent variable from independent variables. The logistic model (or logit version) is used to version the probability of an optimistic elegance or event present along with bypass/fail, win/lose, alive/lifeless, or healthy/sick. This will be prolonged to model numerous classes of occasions, determining whether or not a photo incorporates a cat, canine, lion, *etc.* Every item detected within the image could be assigned a chance between zero and one, and the sum added to one. Logistic Regression is a popular machine learning algorithm, which gives reasonable probabilities and classifies new data using both continuous and discrete datasets. It can classify observations using various data types and quickly determine the most influential variables for classification.

By applying this algorithm model to our system, we got an accuracy of 96.36 percent.

Decision Tree

A decision tree is a decision-making tool that uses a tree-like model of decisions and potential outcomes, such as chance event outcomes, resource costs, and utility. It is one way to represent an algorithm that is entirely composed of conditional control statements. Each internal node represents a "test" on an attribute (for example, whether a coin flip results in heads or tails), each branch represents the result of the test, and each leaf node represents a class label. From root to leaf, the pathways define the classification rules. The algorithm in a decision tree starts at the root node and works its way up to predict the class of a given dataset. Based on the comparison, this algorithm follows the branch and jumps to the next node based on the values of the root attribute and the importance of the record (real dataset) attribute. The algorithm compares the attribute value to the values of the other sub-nodes before proceeding to the next node. It repeats the process until it reaches the leaf node of the tree, as shown in Fig. (**1**).

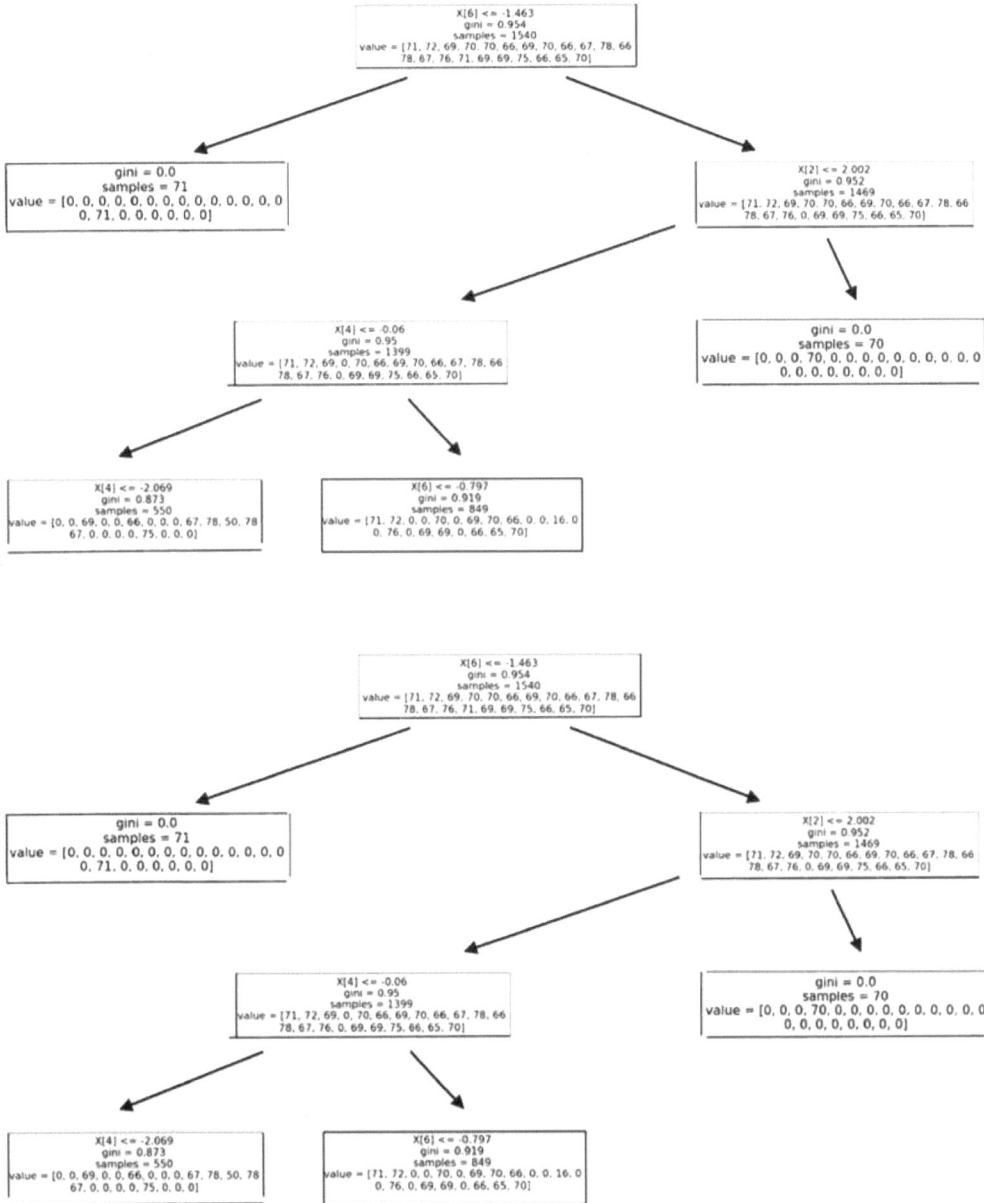

Fig. (1). Flow regimes and governing equations.

By applying this algorithm model to our system, we got an accuracy of 99.54 percent.

k-nearest Neighbours (KNN) Algorithm

The KNN algorithm is a supervised machine learning algorithm that can be applied to classification and regression problems. It is simple to understand and also to set up, but it becomes considerably slower as the amount of data given as dataset grows. KNN works by calculating the minimum distances between a query and all instances in the data, selecting the k examples that are closest to the question, and then selecting the most frequent label (during classification) or averaging the titles (during regression). KNN is mostly used for classification problems as it classifies the new data point based on similarities with existing data points, as shown in Fig. (**2**).

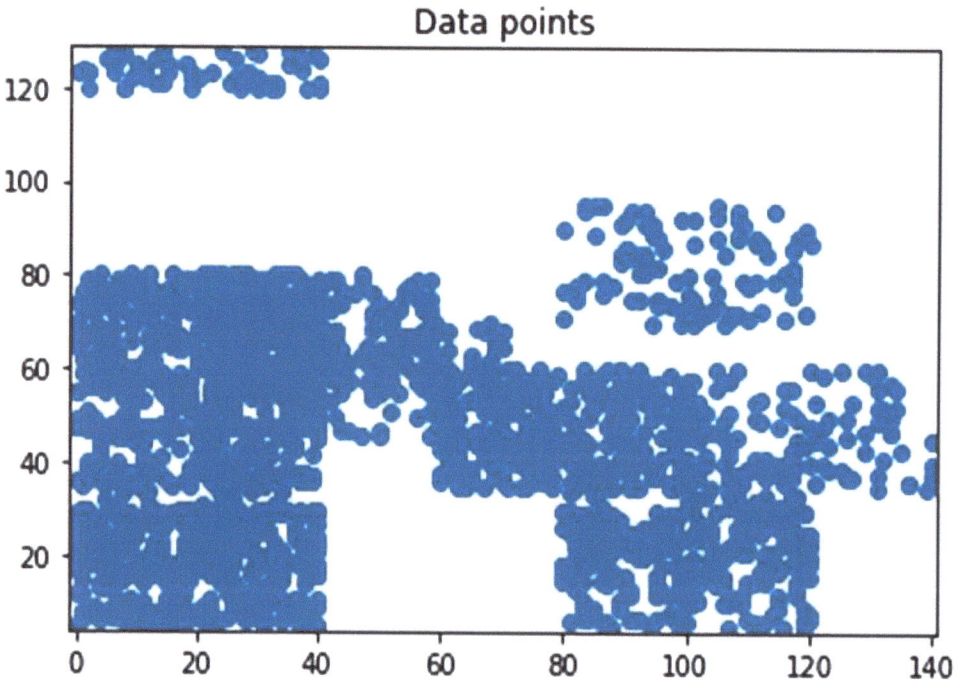

Fig. (2). KNN datapoints of crop recommendation.

By applying this algorithm model to our system, we got an accuracy of 98.03 percent.

Naive Bayes Algorithm

The Naive Bayes algorithm is a supervised learning algorithm that is used to solve classification problems and is based on Bayes' theorem. It is one of the simplest and most effective classification algorithms, assisting in the development of fast machine learning models which are capable of making quick predictions. It is most commonly used in text classification with high-dimensional training datasets. This is a probabilistic classifier, which means it makes predictions based on an object's probability.

By applying this algorithm model to our system, we got an accuracy of 99.09 percent.

After applying all the above algorithms, it is found that decision tree was predicting the highest accuracy with 99.54 percent, as shown in Table **1**.

Table 1. Accuracy of algorithms.

Name of the algorithm	Accuracy attained
Logistic Regression	96.36%
Decision tree	99.54%
k-nearest neighbours (KNN) algorithm	98.03%
Naive Bayes algorithm	99.09%
Logistic Regression	96.36%

IMPLEMENTATION

Data Pre-processing

Data cleaning is performed primarily as part of data pre-processing to clean the data by filling missing values, smoothing noisy data, resolving inconsistencies, and removing outliers. As the data set used does not contain any null values, we don't have to use any missing techniques like fillna() or filling it with mean/median. In the dataset, we have 7 independent features and 1 dependent feature. That one dependent column or Predicted column is Crop names like rice, wheat, *etc.* So we will be using Label encoding to encode the categorical value.

It is important in machine learning programs to distinguish the feature matrix (independent variables) and dependent variables from the dataset. Our dataset contains seven independent variables: nitrogen, phosphorous, potassium, temperature, humidity, pH, and rainfall, as well as one dependent variable: crop name. To extract a variable, we will use the Pandas library's "iloc[]" method. Its function is to extract the required rows and columns from the dataset. After completing the preceding steps, we divide our dataset into two parts: training and testing, as it's an important step in data pre-processing. The training data is around 70% of the data, *i.e.,* 1540, and the test data is approximately 30% of the dataset, *i.e.,* 660. The final step in data pre-processing is feature scaling. It is a method for standardising a dataset's independent variables within a given range, as shown in Fig. (**3**).

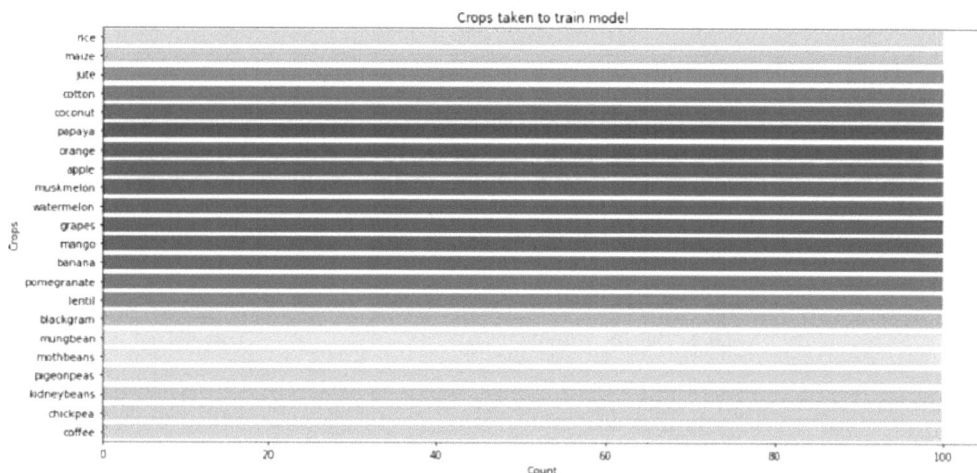

Fig. (3). Crops considered training the given model.

Applying Machine Learning Algorithms

In summary, predictive modelling is a statistical technique that employs machine learning and data mining to predict and forecast likely future outcomes based on historical and existing data. It works by analysing current and historical data and then projecting what it learns onto a model designed to predict likely outcomes.

By applying a few algorithms like Logistic Regression, Decision tree, k-nearest neighbours (KNN) algorithm and Naive Bayes algorithms on the pre-processed dataset, it is found that decision tree was predicting the highest accuracy with 0.45 percent more than Naive Bayes algorithm, as shown in the Fig. (**4**).

Application

Now that the dataset has been pre-processed and the algorithms have been applied, the system's back-end has been completed. We can now start working on our main application (*i.e.,* Front-end of the system). Here, Streamlit - a Python web framework for creating web apps for Machine Learning and Data Science is used. Streamlit is making life easy because it combines the back-end and the front-end of the app.

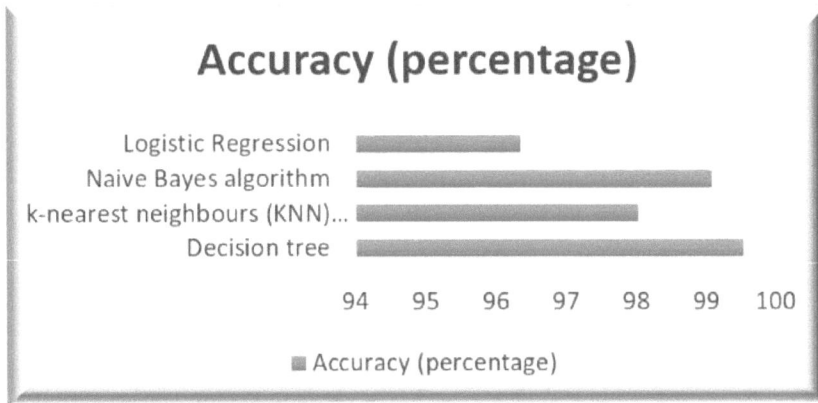

Accuracy (percentage)

Fig. (4). Accuracy of the models (percentage).

The first step to begin building the app is to install the libraries. In addition to Streamlit, we will also be using Pandas to load our data, joblib to provide lightweight pipelining and scikit-learn to implement our machine learning model. Pickling the model is the next step. Pickling is simply the process of saving the model as a .pkl file. The goal is to keep the training process separate from the user experience. "clf = joblib.load("clf.pkl")" is calling the model we trained and pickled earlier. Fetching the value of the input can be done in one file by simply writing, " x = st.number_input ("placeholder")" and the input value is stored directly in the variable x.

RESULTS AND DISCUSSION

By entering the parameters, the values are added into the system as input, and the system analyses the data and returns crop results as output, as shown in Figs. (**5a & b**). The system recommended that crops are grown efficiently so that farmers could maximise production, profit and indicates the crop that is best suited for growing exponentially in soil with soil parameters similar to those entered by the user. The recommended crops for different types of soil conditions, temperature, rainfall, *etc.,* are tabulated in Table **2**.

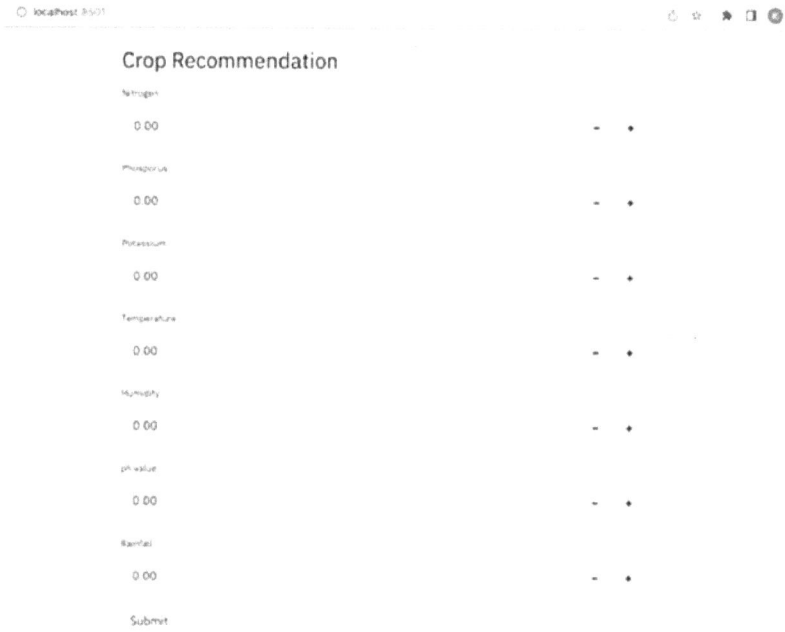

Fig. (5a). Webpage before entering the values.

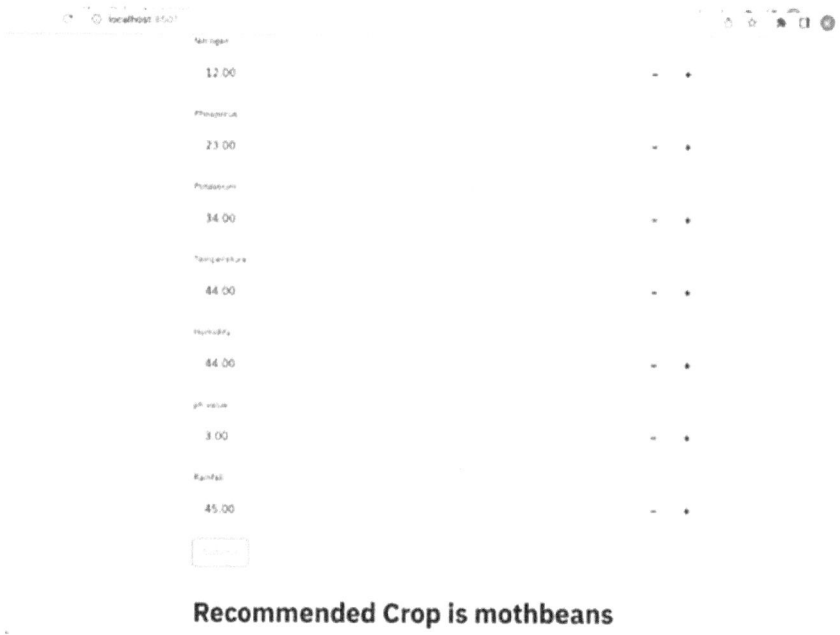

Recommended Crop is mothbeans

Fig. (5b). Webpage after recommending the crop.

Table 2. Recommended crop by Machine learning model.

S.no	Nitrogen	Phosphorus	Potassium	Temperature	Humidity	ph value	Rainfall	Recommended Crop
1	119	44	33	44	33	8	99	Black gram
2	9	564	983	84	66	7	199	Grapes
3	731	815	902	486	123	5	569	Apple
4	917	135	276	687	39	5	9	Muskmelon
5	77	31	0.03	99	456	6	998	Orange
6	90	42	43	21	82	6.5	203	Rice
7	12	23	34	44	44	3	45	Moth beans

CONCLUSION

In this paper, machine learning based user friendly crop recommendation system is developed which is useful for farmers. Farmers can enter the percentage of Nitrogen, Phosphorous, Potassium of the soil, temperature, pH value and rainfall, based on the inputs machine learning model, which give them the recommended crops. The use of prediction models helps to overcome the unpredictability of favourable crops or other agricultural-related problems.

In terms of future scope, the machine learning model can be integrated with real-time data. A real-time app, which gives local soil health data, temperature and rainfall forecast, groundwater pH value, and water availability throughout the year from irrigation projects, is needed to be developed. The real-time data collected through the app over a period of time help to give accurate crop recommendations to farmers. Thereby we can bring technological innovation to farming, and Government can advise the local farmers about soil health and to follow a specific cropping pattern through advisory boards to minimise reduced productivity and overuse of the limited water resources.

REFERENCES

[1]   Z. Doshi, S. Nadkarni, R. Agrawal, and N.K. Shah, "AgroConsultant: Intelligent crop recommendation system using machine learning algorithms", *2018 Fourth International Conference on Computing Communication Control and Automation (ICCUBEA)*, pp. 1-6, 2018.
[http://dx.doi.org/10.1109/ICCUBEA.2018.8697349]

[2]   S. Pudumalar, E. Ramanujam, R.H. Rajashree, C. Kavya, T. Kiruthika, and J. Nisha, "Crop recommendation system for precision agriculture", *2016 Eighth International Conference on Advanced Computing (ICoAC)*, pp. 32-36, 2017.
[http://dx.doi.org/10.1109/ICoAC.2017.7951740]

[3]   M.J. Mokarrama, and M.S. Arefin, "RSF: A recommendation system for farmers", *2017 IEEE Region 10 Humanitarian Technology Conference (R10-HTC)*, pp. 843-850, 2017.

[4]   S.M. Pande, P.K. Ramesh, P. Anmol, B.R. Aishwarya, K. Rohilla, and K. Shaurya, "Crop recommender system using machine learning approach", *2021 5th International Conference on Computing Methodologies and Communication (ICCMC),* pp. 1066-1071, 2021.

[5]   A. P. Chakraborty, S. Kumar, and O.R. Pooniwala, "Intelligent crop recommendation system using machine learning", *2021 5th International Conference on Computing Methodologies and Communication (ICCMC),* pp. 843-848, 2021.

[6]   A. Kumar, S. Sarkar, and C. Pradhan, "Recommendation system for crop identification and pest control technique in agriculture", *2019 International Conference on Communication and Signal Processing (ICCSP),* pp. 0185-0189, 2019.
        [http://dx.doi.org/10.1109/ICCSP.2019.8698099]

[7]   D. Modi, A.V. Sutagundar, V. Yalavigi, and A. Aravatagimath, "Crop recommendation using machine learning algorithm", *2021 5th International Conference on Information Systems and Computer Networks (ISCON),* pp. 1-5, 2021.
        [http://dx.doi.org/10.1109/ISCON52037.2021.9702392]

CHAPTER 12

Artificial Neural Networks based Distributed Approach for Heart Disease Prediction

Thakur Santosh[1,*]**, Hemachandran K.**[4]**, Sandip K. Chourasiya**[3]**, Prathyusha Pujari**[2]**, K. Vishal**[2] **and B. R. S. S. Sowjanya**[2]

[1] *Mahindra University, Hyderabad, India*

[2] *Woxsen School of Business, Woxsen University, Kamkole, Sadasivpet, Telangana, India*

[3] *University of Petroleum and Energy Studies, Dehradun, India*

[4] *Department of Artificial Intelligence, School of Business, Woxsen University, Hyderabad, India*

Abstract: A recent study shows that almost 30% of total global deaths are caused by heart disease. These days precise diagnosis related to heart disease is very difficult. The doctor advises patients to take various tests for diagnosis, which is a very costly and time-consuming process as medical databases are large and cannot be processed quickly. A new approach has been proposed to predict heart disease from historical data sets. In this chapter, heart disease possibilities in patients are predicted with the help of neural networks on distributed computing. Feature selection was applied to the dataset to get better results and to increase the performance. Feature selection reduces the number of attributes from the dataset and only provides the necessary attributes, which directly reduces the number of tests required for the diagnosis.

Keywords: Artificial Neural Networks (ANN), Distributed Computing, Hadoop, Hadoop Distributed File System (HDFS).

INTRODUCTION

Today, heart disease is one of the leading causes of death in many countries due to changes in lifestyle, food habits and smoking and alcohol consumption. Different health activities is also one of the reasons for heart diseases like hypertension, obesity, *etc*. Therefore, more efficient methods are required to predict heart diseases [1]. Today, cardiovascular diseases, primarily heart disease, are diagnosed by an expert doctor. Even though some cases are wrongly diagnosed, patients are advised to undergo many tests for precise diagnosis, which

---

[*] **Corresponding author Thakur Santosh:** Mahindra University, Hyderabad, India;
E-mail: Santosh.thakur@mahindrauniversity.edu.in

Hemachandran K., Raul V. Rodriguez, Umashankar Subramaniam & Valentina Emilia Balas (Eds.)

is very costly, and large medical databases cannot be processed quickly. It also makes the diagnosis process very time-consuming. Therefore, data analytics has become a basic need of the medical healthcare world.

Data analytics is the process of extracting valuable information from a large amount of databases. It is a very important method in the medical field where it extracts biomedical and healthcare knowledge and provides a great help in complicated medical decisions [2]. Pre-processing the data is a vital role in data analysis. In this process, the unwanted data is removed refined. It also enhances the quality of diagnosis results.

Unprocessed data can increase the computation time and can affect the accuracy of the results. This type of useless data can be eliminated before learning with the help of feature selection. Feature selection is the process to find a minimum set of attributes that are necessary for the current operation and eliminate irrelevant, redundant, or noisy data [3]. This reduction gives a great impact on the data mining process and also increases the accuracy level of diagnosis. ANN is an information processing paradigm which directly stimulates the human brain and can be applied to various real world problems. Neural networks are well trained for pattern recognition, and also for storing and retrieving patterns to solve combinatorial optimization problems. These abilities also make neural networks very good for classification problems. When there is the use of high-dimensional datasets, we arise a problem called data dimensionality, which cannot be handled with traditional machine learning algorithms. When we perform iterations with machine learning algorithms, it consumes a large amount of time and leads to low accuracy. Hence to improve the accuracy of the model and to implement machine learning algorithms. A novel distributed approach is implemented.

In this chapter, Linear Discernment Analysis (LDA) feature selection and Distributed Artificial Neural Networks (D-ANN) are applied to classify the instances. The main aim of this chapter is to predict heart disease in a patient with reduced number of attributes, which directly reduces various numbers of tests required to diagnose the presence of heart disease.

This chapter is organized as follows: Section 1 provides an introduction to the work. Section 2 proposes related work. Section 3 explains the overview of methods that are used in our research. Section 4 describes the experimental setup and data set. Section 5 provides the experiments and results. Section 6 gives the conclusion.

RELATED WORK

Few researchers have done some interesting work for the diagnosis of different types of heart diseases. Our approach is to apply the ANN to distributed computing for the prediction of heart disease. Patil *et al.* [4] presented a heart disease prediction model. Here they applied a multilayer neural network. In this model, 13 clinical features are used as input for ANN and a back-propagation algorithm is used for training purposes.

Deekshatulu *et al.* [2] applied feature selection with ANN and performed the classification of various cardiovascular diseases. For the data pre-processing, PCA is applied to reduce the number of attributes. Suganya *et al.* [4] presented a neural model which can predict heart disease in an early stage with a minimum number of attributes. In this model, multilayer neural networks are applied. Subanya *et al.* [5] applied a metaheuristic algorithm to detect heart disease. They used a Binary Artificial Bee Colony (BABC) algorithm to regulate the set of optimal numbers of features with better accuracy. Finally, the KNN method is applied for classification. Priti Chandra *et al.* [6] presented a new approach to heart disease prediction. They used hybrid feature selection with associative classification. In this method, they pruned irrelevant, redundant attributes from the given data and produced a solid rule set. These rules are written by the physicians to build the classifier for the detection of heart disease in a patient. M. Anbarsai *et al.* [1] applied Genetic Algorithm (GA) to detect heart disease with a minimum number of attributes. To classify the instances, Naive Bayes and Decision Tree are applied. Yasin Kaya *et al.* [7] focused on the classification of heartbeats from ECG signals. Selected features were used for performance evaluation. The main objective is to use GA for feature selection to select the best features and integrate them for analysis. To obtain the experimental results, various classification algorithms like neural network, support vector machine and K-nearest neighbor were applied. Training a dataset with noise and irrelevant data may lead to less accuracy for classification. Therefore feature subset selection is used as the data pre-processing step, which plays a significant role in the field of data analysis. Abeer S. Desuky *et al.* [3] proposed a wrapper approach with Particle Swarm Optimization (PSO) for feature selection on large medical databases. The selected features are used as input to five classifiers, and the output is compared with other researcher's work. An intelligent Heart Disease Prediction System (IHDPS) was developed [8]. In this method, they used three data mining techniques to build the model. This model can be used to find and extract hidden knowledge associated with heart disease from a large historical heart-related database. It can also answer heart disease-related complex queries and can help doctors in better decision-making. If proper features of a system are selected, the efficiency and performance of that system will definitely increase in terms of cost, time and accuracy.

Neelakantan *et al.* proposed a method in which a Multi-Layer Neural Network is used as a method for feature selection. This method selects interesting features from an Ischemic Heart Disease (IHD) database. This method reduced the number of attributes from 17 to 12. The main objective was to observe back propagation neural networks to study IHD [9]. In database classification, feature selection plays an important role. It simply removes insignificant features, which reduces the computational cost and execution time and makes the diagnosis process more easy and accurate.

Jaganathan *et al.* [10] proposed a different feature subset selection method which is based on fuzzy entropy measures and provides a new way to handle the medical database classification. Evaluation of the proposed method is done using a Radial Basis Function (RBF) classification algorithm [10 - 13].

Singh *et al.* [14] proposed a model to reduce the features using discriminant analysis with binary classification. Cheng *et al.* [15] proposed a feature clustering method to select the features and form a cluster. With this feature clustering approach, the number of features is reduced to pass the data into the model. Yang *et al.* [16] proposed fuzzy C-means approach to predict heart disease. In this model, authors apply fuzzy C-means to select the weighted features. Zang *et al.* [17] applied ada boosting and principal component analysis to predict heart disease. In this model, authors applied feature selection to reduce the dimensionality of the data. But when we are using iterative models on high dimensional data. The system can consume more amount of time.

The above methods presented in this chapter for heart disease prediction are based on feature selection or feature reduction strategies. But these methods are not explored in the dimension of a distributed approach. When we run computationally complex algorithms on high dimensional data, it leads to different problems, as mentioned above.

Hence in this chapter, we propose a new distributed approach to predict heart disease on clinical parameters, *i.e.,* ANN based distributed computing.

The MonkeypoxHybridNet is a hybrid deep convolutional neural network model for monkeypox disease detection. It is used to detect the virus that causes monkeypox, which is found in the bodies of rodents and other animals and is passed from infected animals to people. A contemporary technique in recent years has been image processing using artificial intelligence. Using deep convolutional neural networks to infer inferences from images and photographs (CNN). For various disorders, this approach is used. There are numerous techniques to create deep learning networks (Kurmi *et al.*, 2022). Here, our goal is to successfully create a deep CNN for monkeypox detection.

The monkeypox virus is now highly widespread. Many fear it due to their COVID-19 encounters. Monkeypox sickness can be identified using deep CNNs. MonkeypoxHybridNet, a brand-new deep CNN model, is suggested in this paper. ResNet50, VGG19, and InceptionV3 models make up this dataset. The four classes in the dataset are used to train the proposed model. The four categories include: normal, measles, chickenpox, and monkeypox. The suggested model outperforms the alternative models. The accuracy values for the ResNet, VGG19, and InceptionV3 models are 0.595, 0.705, and 0.805, respectively. Furthermore, MonkeypoxHybridNet reports an accuracy of 0.842. It demonstrates the effectiveness of MonkeypoxHybridNet in identifying monkeypox disease from skin scans. New models or novel hybrid algorithms for monkeypox identification can be suggested for future research.

OVERVIEW OF THE MODEL

As presented in (Fig. **1**), ANN is composed of multiple nodes, which simulate the biological neurons of the human brain. The ANN is interconnected and they can easily interact with each other. The data can be fed to these nodes and passed through the connected nodes for output. The output from each node is passed through the activation function, and finally, the error is calculated. In clinical diagnosis, patients risk factors are given as input to the network.

An artificial neural network consists of three simple layers, which are input layer, hidden layer and output layer. It is configured for specific applications, such as pattern recognition or data classification, through a learning process. These networks are very reliable for solving real-world problems. In a neural network, we don't tell the machine how to solve our problems. It learns from observation data, finds patterns and figures out its own solution to the problem.

EXPERIMENTAL SETUP

To perform extensive experiments, Spark 1.6.0 is installed on top of the Hadoop 2.6.0 version. With the host operating system as UBUNTU 14.04, the methodology is examined on three nodes.

Data set description: The datasets considered for this work contains 13 attributes with 270 instances, and these data sets are augmented in multi folds to place in distributed computing.

Fig. (1). Representation of simple neural networks.

Age: represents the age of the patient.

Sex: represents male or female.

Chest pain type: type of chest pain.

Resting blood pressure: normal blood pressure with systole and diastole movement.

Serum in cholesterol in mg/l: It shows the amount of triglycerides present in the blood.

Blood sugar: fasting blood sugar greater than 125 mg/dl (1 true) is diabetes, less than 100 mg/dL (5.6 mmol/L) is normal, and 100 to 125 mg/dL (5.6 to 6.9 mmol/L) is considered prediabetes.

ECG: resting electrocardiographic results.

Maximum heart rate: maximum heart rate achieved. The maximum heart rate is 220 – your age.

Angina: it is the type of pain occurred in the chest for minimum blood flow in the heart.

Old peak: ST depression induced by exercise relative to rest (ST represents the positions on ECG plot).

Slope peak excersie: the slope of ST segment.

Major vessels: number of major vessels colored.

Thal: it represents the blood disorder called thalassemia value.

RESULTS

The heart disease dataset was then divided into 2 categories, *i.e.,* training data and testing data. Training data consists of a total of 199 tuples, and testing data consists of 71 tuples. Then a neural network algorithm was applied to this dataset. As a result of which, the error rate came to 12.69, which was high. It shows that the performance of the model is very low. The error rate is high because the data set may have some irrelevant and redundant features, which may decrease the performance. To increase the performance, a feature selection method is applied to reduce the number of attributes. The reduced features are presented in Fig. (**2**).

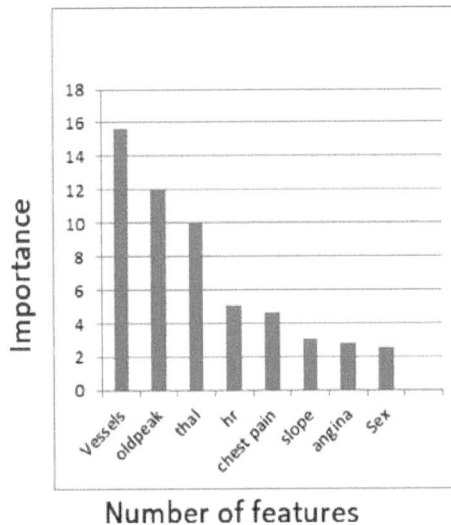

Fig. (2). Reduced number of features and its importance.

The comparison of D-ANN with different algorithms is given in Tables **1a** & **b** and Figs. (**3** & **4**). The D-ANN algorithm achieves an accuracy of 93.57% and consumes less time.

Table 1a. System configuration.

-	Master	Node1	Node2
CPU Configuration	ML 350E, Gen 8, Intel Xeon CPU E5-24070@2.20 GHz	Intel, Core i7-4510UCPU @ 2GHz	Intel, Core i7-4510U CPU @ 2GHz
RAM	16GB	8GB	8GB

Data nodes and Name nodes are the two interior components of a Spark cluster. In a cluster, there exists more than one data node and only a single name node. The work of the name node is to assign jobs to all the data nodes. A heart disease dataset from UCI machine learning repository [12, 13] has been used in this work. The data set contains thirteen attributes. The label attribute is the class identifier with the value "0" indicating no heart disease and value "1" indicating the presence of heart disease.

Table 1b. Comparison of different algorithms with proposed work.

Classification Algorithm	Execution Time	Accuracy
D-ANN	21	93.57
Decision trees	65	92.1
Random Forest	62	91.5
ANN	61	91.2

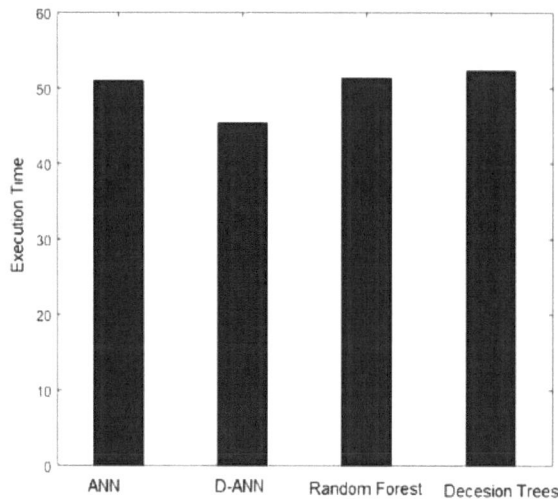

Fig. (3). Execution time in comparison with other algorithms.

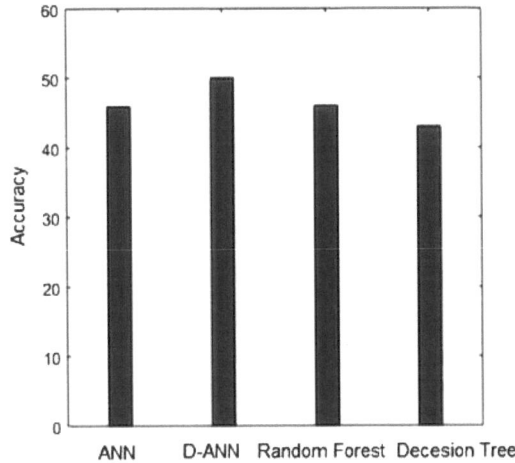

Fig. (4). Accuracy in comparison with other algorithms.

CONCLUSION

Our research aims to identify the existence of cardiac disease using the fewest possible variables. Here, we used linear discriminant analysis (LDA) to apply feature selection and then distributed artificial neural network (D-ANN) to classify the data. The system's computation time is decreased and good accuracy is attained with the help of the suggested distributed technique. The model's training time was also shortened with this approach. We anticipate using this D-ANN model in the future to analyse massive amounts of data with greater dimensionality. This suggested model will not only be tested on one dataset but also on others, and we will work to increase the model's accuracy before putting it into use.

This work will be expanded in order to handle massive datasets more quickly and perform better. The distributed technique shortens the training period and assists in resolving the issue of handling enormous data. Real-time applications may be appropriate for this architecture. To demonstrate our model's effectiveness, we would like to test it on several other medical datasets.

Our objective is to further develop this D-ANN model to quickly and accurately predict outcomes from large datasets with high dimensionality. To show how effective it is, the proposed model should be evaluated on several different datasets in addition to just one. We want to improve this work so that it can handle incredibly huge datasets quickly and perform better. The distributed technique reduces training time while helping to solve the issue of processing massive volumes of data. It is possible that this paradigm will work well in real-time applications. To show this model's effectiveness, we want to test it against numerous medical datasets.The goal of this research is to create a system that can identify the presence of heart disease using just a few variables. We used Linear Discriminant Analysis (LDA) to do feature selection before implementing Distributed Artificial Neural Network (D-ANN) for classification.

REFERENCES

[1] M. Anbarasi, E. Anupriya, and N.C.S.N. Iyengar, "Enhanced prediction of heart disease with feature subset selection using genetic algorithm", *Int. J. Eng. Sci. Technol.*, vol. 2, no. 10, pp. 5370-5376, 2010.

[2] B. L. Deekshatulu, "Classification of heart disease using artificial neural networks and feature subset selection", *Global Journal of Computer Science and Technology.*, vol. 13, no. 3, 2013.

[3] H.M. Harb, and A.S. Desuky, "Feature selection on classification of medical datasets based on particle swarm optimization", *Int. J. Comput. Appl.*, vol. 104, p. 5, 2014.
[http://dx.doi.org/10.5120/18197-9118]

[4] J.S. Sonawane, and D.R. Patil, "Prediction of heart disease using a multilayer perceptron neural network", *Information Communication and Embedded Systems (ICICES), 2014 International Conference on. IEEE*, 2014.
[http://dx.doi.org/10.1109/ICICES.2014.7033860]

[5] B. Subanya, and R. R. Rajalaxmi, "A novel feature selection algorithm for heart disease classification", *International Journal of Computational Intelligence and Informatic*, vol. 4, no. 2, 2014.

[6] P. Chandra, and B.L. Deekshatulu, "Prediction of risk score for heart disease using associative classification and hybrid feature subset selection", *Intelligent Systems Design and Applications (ISDA), 2012 12th International Conference on. IEEE*, 2012.
[http://dx.doi.org/10.1109/ISDA.2012.6416610]

[7] Y. Kaya, and H. Pehlivan, "Feature selection using genetic algorithms for premature ventricular contraction classification", *Electrical and Electronics Engineering (ELECO), 2015 9th International Conference on. IEEE*, 2015.
[http://dx.doi.org/10.1109/ELECO.2015.7394628]

[8] S. Palaniappan, and R. Awang, "Intelligent heart disease prediction system using data mining techniques", *Computer Systems and Applications, 2008. AICCSA 2008. IEEE/ACS International Conference on. IEEE*, 2008.
[http://dx.doi.org/10.1109/AICCSA.2008.4493524]

[9] K. Rajeswari, V. Vaithiyanathan, and T.R. Neelakantan, "Feature selection in ischemic heart disease identification using feed forward neural networks", *Procedia Eng.*, vol. 41, pp. 1818-1823, 2012.
[http://dx.doi.org/10.1016/j.proeng.2012.08.109]

[10] P. Jaganathan, and R. Kuppuchamy, "A threshold fuzzy entropy based feature selection for medical database classification", *Comput. Biol. Med.*, vol. 43, no. 12, pp. 2222-2229, 2013.
[http://dx.doi.org/10.1016/j.compbiomed.2013.10.016] [PMID: 24290939]

[11] B.L. Ramani, P. Poosapati, P. Tumuluru, C.H.M.H. Saibaba, M. Radha, and K. Prasuna, "Deep learning and fuzzy rule-based hybrid fusion model for data classification", *International Journal of Recent Technology and Engineering*, vol. 8, no. 2, pp. 2277-3878, 2019.
[http://dx.doi.org/10.35940/ijrte.B2304.078219]

[12] K. Verma, S. Bhardwaj, R. Arya, M.S.U. Islam, M. Bhushan, A. Kumar, and P. Samant, "Latest tools for data mining and machine learning", *Int. J. Inno. Tech. Exp. Engi.*, vol. 8, no. 9S, pp. 18-23, 2019.

[http://dx.doi.org/10.35940/ijitee.I1003.0789S19]

[13] D. Zhang, L. Zou, X. Zhou, and F. He, "Integrating feature selection and feature extraction methods with deep learning to predict clinical outcome of breast cancer", *IEEE Access,* vol. 6, pp. 28936-28944, 2018.
[http://dx.doi.org/10.1109/ACCESS.2018.2837654]

[14] R.S. Singh, B.S. Saini, and R.K. Sunkaria, "Detection of coronary artery disease by reduced features and extreme learning machine", *Clujul Med.,* vol. 91, no. 2, pp. 166-175, 2018.
[PMID: 29785154]

[15] R. Chen, N. Sun, X. Chen, M. Yang, and Q. Wu, "Supervised feature selection with a stratified feature weighting method", *IEEE Access,* vol. 6, pp. 15087-15098, 2018.
[http://dx.doi.org/10.1109/ACCESS.2018.2815606]

[16] M.S. Yang, and Y. Nataliani, "A feature-reduction fuzzy clustering algorithm based on feature-weighted entropy", *IEEE Trans. Fuzzy Syst.,* vol. 26, no. 2, pp. 817-835, 2018.
[http://dx.doi.org/10.1109/TFUZZ.2017.2692203]

[17] R. Zhang, S. Ma, L. Shanahan, J. Munroe, S. Horn, and S. Speedie, Automatic methods to extract New York heart association classification from clinical notes.*ieee international conference on bioinformatics and biomedicine (bibm)* IEEE, 2022, pp. 1296-1299.

CHAPTER 13

Reinforcement Learning Based Automated Path Planning in Garden Environment using Depth - RAPiG-D

S. Sathiya Murthi[1,*], **Pranav Balakrishnan**[1], **C. Roshan Abraham**[1] and **V. Sathiesh Kumar**[1]

[1] *Madras Institute of Technology, Anna University, Chennai, India*

Abstract: Path planning by employing Reinforcement Learning is a versatile implementation that can account for the ability of a robot to autonomously map any unknown environment. In this paper, such a hardware implementation is proposed and tested by making use of the SARSA algorithm for path planning and by utilizing stereovision for depth estimation based obstacle detection. The robot is tested in a cell-based environment – 3x3 with 2 obstacles. The goal is to map the environment by detecting and mapping the obstacles and finding the ideal route to the destination. The robot starts at one end of the environment and runs through it for a specified number of episodes, and it is observed that the robot can accurately identify and map obstacles and find the shortest path to the destination in under 10 episodes. Currently, the destination is a fixed point and is taken as the other diagonal end of the environment.

Keywords: Adaptive, Autonomous, Cell-based, Closed environment, Depth Estimation, Depth map, Dynamic, Episodes, Map, Micro-controller, Obstacle Detection, On Policy, Path planning, Q-Table, Reinforcement learning, Robot, Route, SARSA, Stereo Vision, Thresholding.

INTRODUCTION

Reinforcement Learning is an adaptable and learning-based method that can allow an agent to learn and determine the optimal solution to a problem on its own based on the rewards and penalties offered by the environment [1]. This allows the agent to handle any state that it might encounter while learning to achieve the required results. SARSA or State-Action-Reward-State-Action is a basic Reinforcement Learning Algorithm, an On Policy technique. With increasing innovations and implementations of Reinforcement Learning in Self-driving vehicles [2], Autonomous Robots [3], and UAVs [4, 5], there rises a requirement

* **Corresponding author S. Sathiya Murthi:** Madras Institute of Technology, Anna University, Chennai, India; E-mail: sathiyamurthi239@gmail.com

for adaptive mapping of unknown environments and the ability to determine the shortest path to any given destination to handle unpredictable and dynamic environments. Reinforcement Learning is an effective method that can be used to satisfy the requirements of such an environment by learning from repeated iterations of trials to estimate the optimal path from source to destination while negotiating obstacles [6]. In this paper, this implementation has been employed to plan routes and identify obstacles in a garden environment which can be used to automate the management and caretaking of gardens and plants. This technique can further be extended to exploring and searching tasks such as space exploration and mapping, Air crash and accident investigations.

RELATED WORKS

To discover the path between source and destination, Konor *et al.* reported on an enhanced Q-learning approach [7]. The step distance (from one state to the next) and the eventual destination are assumed here. It is used to update the entries in the Q-table. Unlike the traditional Q-learning approach, where the values are continually updated, the values are only entered once. At each state, the Q-value derived for the best action is saved. In terms of traversal time and the number of states traversed, performance tends to increase [8]. describes end-to-end path planning using Deep Reinforcement Learning. To estimate the Q-value for each state-action, a deep Q-network (DQN) is first created and trained. The RGB picture frame is fed into the DQN. The best course of action is chosen using an action selection approach. The authors claimed that using the DQN approach for path planning resulted in a successful outcome. Path planning is done using a Q-learning algorithm based on the Markov Decision Process [9], according to Sichkar *et al.* [10].

It is challenging to find an optimal path in complicated situations using the traditional Q-learning approach. The robot determined/identified an optimal path from the source to the destination by avoiding collisions with impediments in its propagation path, according to the authors. The shortest path between the source and destination is determined using Q-learning and SARSA algorithms [11]. The method has been tested in a simulated environment with preset barriers. Different learning periods are included in the two algorithms used. It also fluctuates in the number of steps it takes to get to its objective by avoiding collisions with objects along the route.

The shortest path between the source and destination cannot be found using traditional Breadth First Search (BFS) or Rapidly Exploring Random Trees (RRT) techniques. As a result, the authors designed and showed a path-planning algorithm based on reinforcement learning [12]. To begin, a random route graph is chosen. If the chosen path has barriers, it is not taken into account. A collision-free route is found using the Q-learning approach. When compared to RRT and BFS algorithms, the suggested approach provided a smooth and quickest path, according to the authors.

In an unknown environment, the iterative SARSA algorithm [13] is used to discover the best path from the source to the destination. Traditional Reinforcement learning techniques are contrasted on criteria like route length and processing complexity (Q-learning and SARSA). The authors claim that as compared to typical Reinforcement learning approaches, the Iterative SARSA algorithm used during robot path planning produces better results.

Based on a thorough review of the literature, it has been determined that path-planning Reinforcement learning is still in its infancy. The algorithm may be fine-tuned or improved further so that it can be used in real-time situations. The use of the Reinforcement learning algorithm in connection to path planning in an unknown environment is investigated in this work. In the process of picking a suitable action for the robot, the SARSA algorithm is applied.

METHODOLOGY

Reinforcement learning is a machine learning technique based on rewarding desired behaviors and/or penalizing undesired actions. A reinforcement learning agent can observe and interpret its environment, take actions and learn through experience.

Fig. (1) depicts the basic flow of the SARSA algorithm: State – Action – Reward – State – Action. A software implementation of the SARSA algorithm was initially tested to check the validity of the path-planning algorithm and its ability to map an unknown environment. The algorithm is then modified to make it suitable for hardware implementation.

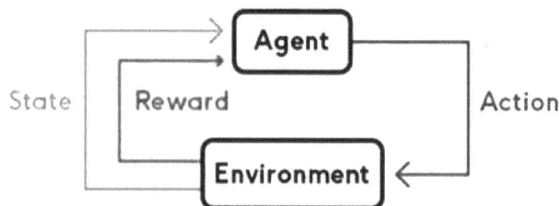

Fig. (1). SARSA Illustration.

The SARSA algorithm involves learning the environment by choosing actions at each state using a policy function. The policy function used in this implementation is that ninety percent of the actions are chosen to maximize reward based on Q-values of the state, and ten percent of the actions are taken at random. At each state, an action is chosen, and then the environment provides the reward for that action along with the next state. Based on the reward, the Q-values are updated in the Q-table. An episode ends when the agent reaches the destination or detects an obstacle. Until then, the agent explores the environment. By doing this, multiple iterations of this episode allow the agent to identify the optimal path to reach the destination by assigning a positive reward for the destination and a negative reward for obstacles.

Stereo vision is used for the computation of depth based on the binocular disparity between the images of an object in the left and right eyes. The Kinect camera, shown in Fig. (**2**), is a motion-sensing input device produced by Microsoft and released in 2010. The device generally contains RGB cameras, infrared projectors, and detectors that map depth through structured light and time of flight calculations, which can, in turn, be used to perform real-time gesture recognition and body skeletal detection, among other capabilities. Image processing techniques are used to extract and utilize the depth map generated by the Kinect camera.

Fig. (2). XBOX 360 Kinect camera.

To perform hardware implementation, the NVIDIA Jetson Nano Developer Kit is used as the Micro-controller to run the developed algorithm and interface with the XBOX 360 Kinect camera, which is shown in Fig. (**2**), and the required sensors for calibrated movement from one cell to the next in the environment. The Kinect camera provides the required depth map by utilizing a stereo camera setup along with an IR camera. An MPU6050 sensor is used to turn accurately by integrating the angular velocity output of the sensor to obtain angular displacement. A Li-Po battery is used to power four 12V 100 RPM DC motors. The Jetson Nano is externally powered using a 5V4A AC adapter.

Fig. (**3**) depicts the workflow of the algorithm implemented for each episode. Fig. (**4**) shows the decision-making process at each cell. Fig. (**5**) shows the algorithm for hardware implementation of path planning. At each state, *i.e.,* at each cell, the agent – the robot chooses an action and then takes the step if it is valid based on the action by turning to the required direction – up, down, left or right. The step involves checking whether the next state has an obstacle using the depth map generated by stereo camera depth estimation. The obstacle detection is optimized by applying a threshold on the depth map generated to identify objects only in the immediate next cell. Additionally, a region of interest is also applied to detect obstacles accurately. The agent then checks if the next state is an empty space, an obstacle, or the destination. This is then used to populate the map of the environment.

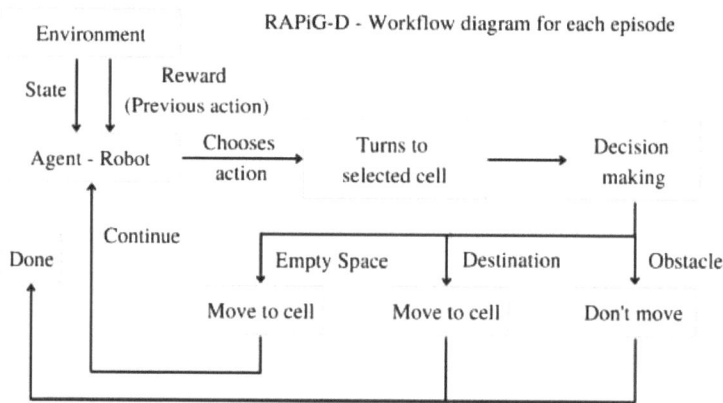

Fig. (3). Flow diagram for each episode.

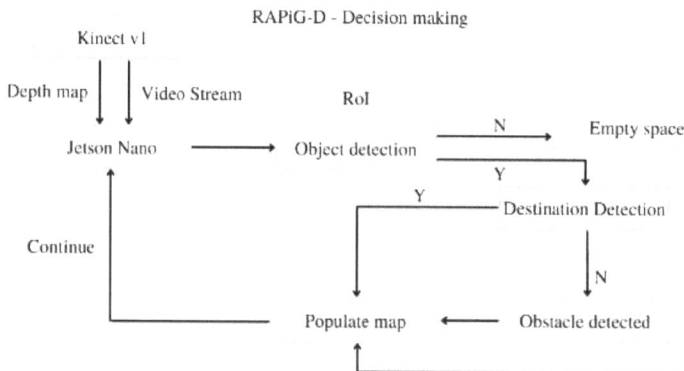

Fig. (4). Flow diagram for decision-making at each cell.

RL Path Planning

Reset Environment ⟶ Initialize parameters ⟶ Render Environment

If obstacle / goal   Else, continue
=> end of episode

Take step	Choose action
- Turn choosen direction	- Add state to table
- Check for obstacle	- Use policy to choose action:
- Move to next state if not obstacle	90% -> best action based on q
- Update map	10% -> random action
- Assign reward	- Check if next state is valid

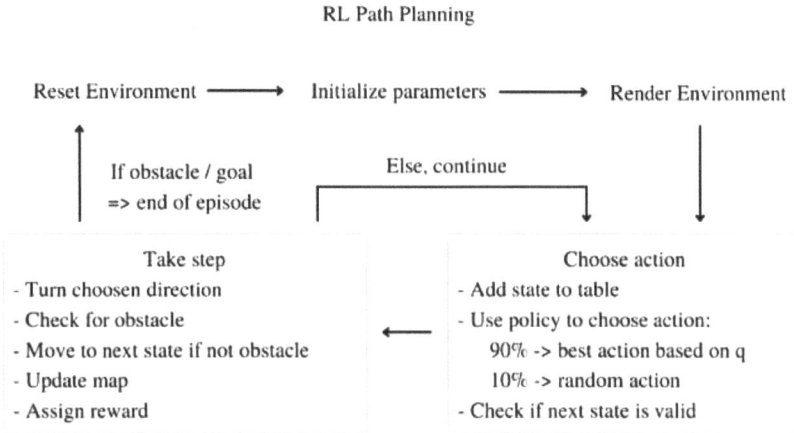

Fig. (5). Flow diagram for Path planning algorithm in hardware implementation.

Figs. (**6** & **7**) show the front and side views of the robot designed to implement the SARSA algorithm as a hardware solution to path planning and to implement obstacle detection using stereo camera-based depth estimation.

Fig. (6). Front view of the robot designed.

Fig. (7). Side view of the robot designed.

Both the software simulation and the hardware implementation follow the same algorithm except for the additional implementation of Obstacle detection using depth estimation in hardware implementation as opposed to feeding obstacle coordinates in the simulation.

A Q-table is initialized with zeros at the start of the first episode, which is used to determine how valuable it is to take a particular action at a particular state to maximize the reward received. The Q-table is updated according to the reward offered for the next state for each action at the current state. A reward of 1 is offered if the next state is the destination, 0 if it is an empty space, and -1 if it is an obstacle. In each episode, the agent goes around the environment, exploring until it reaches either an obstacle or the destination. If the agent reaches the destination, the path from the start point is recorded to check if it is the shortest path. This is used to identify the optimal path to the destination, which can then be used to traverse to the destination.

RESULTS AND DISCUSSION

Software Implementation

Software implementation was performed on a larger 10x10 environment with a larger number of obstacles and was run for 1000 episodes as software implementation takes less time per episode. Fig. (**8**) shows the map generated by the SARSA algorithm when run as a software simulation, and Fig. (**9**) shows the plot of the number of steps taken in each episode. The obstacle coordinates are pre-determined and fed to the algorithm. The Reinforcement Learning agent moves through the environment based on the policy and checks if each co-ordinate is an obstacle, empty space, or destination. In this way, every time it reaches the destination, the shortest path is updated, allowing the agent to identify the optimal path to the destination while avoiding obstacles.

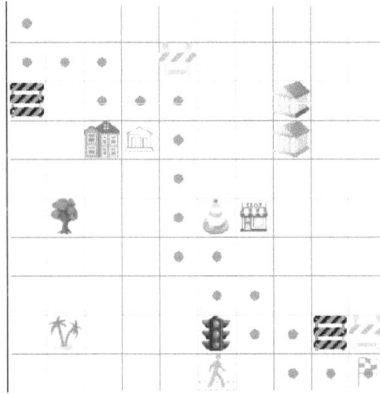

Fig. (8). SARSA – Software implementation- Map of environment.

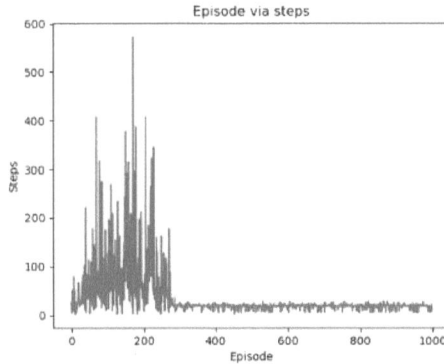

Fig. (9). SARSA – Software implementation- Performance metrics.

Hardware Implementation

The hardware implementation was then performed using a robot controlled by NVIDIA Jetson Nano with an XBOX 360 Kinect camera for depth estimation in a smaller 3x3 environment with two obstacles. The cell-based environment consists of 2 feet x 2 feet cells.

Fig. (10) shows the map generated by the agent as it goes around the environment, exploring and detecting obstacles. The detected obstacles are populated on the map. It also shows the optimal path from start to destination as identified by the agent, thus performing path planning using reinforcement learning.

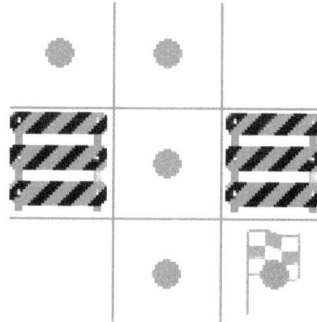

Fig. (10). Map generated for 3x3 environment.

Fig. (**11**) shows the Q-table generated by the RL SARSA algorithm, which shows how valuable each action is at a given state with the states as rows and the actions as columns – Up, Down, Right, and Left.

```
Length of final Q-table = 3
Final Q-table with values from the final route:
                  0        1         2       3
[40.0, 0.0]   0.0  0.000008  0.00000  0.00
[40.0, 40.0]  0.0  0.000882  -0.01000 -0.01
[40.0, 80.0]  0.0  0.000000  0.04901  0.00

Length of full Q-table = 8
Full Q-table:
                  0        1             2       3
[0.0, 0.0]    0.00  -0.010000  3.615840e-08  0.00
obstacle      0.00  0.000000   0.000000e+00  0.00
[40.0, 0.0]   0.00  0.000008   0.000000e+00  0.00
[80.0, 0.0]   0.00  -0.010000  0.000000e+00  0.00
[40.0, 40.0]  0.00  0.000882   -1.000000e-02 -0.01
[40.0, 80.0]  0.00  0.000000   4.900995e-02  0.00
[0.0, 80.0]   -0.01 0.000000   2.673090e-04  0.00
goal          0.00  0.000000   0.000000e+00  0.00
```

Fig. (11). Q-Table for the 3x3 environment – Hardware implementation.

Fig. (**12**) shows the plot of the number of steps in each episode. In a 3x3 environment, the agent finds the optimal path within 10 episodes and traverses to the destination in the optimal path based on the Q-table. As the obstacles and the size of the environment increases, the time required to identify the optimal path would also increase.

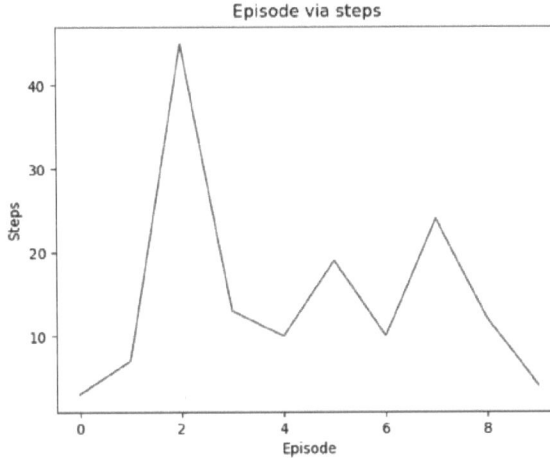

Fig. (12). Plot of the number of steps in each episode – 3x3 environment.

This can be implemented in a garden environment where the robot agent can be tasked to reach a target plant in an optimal path and can monitor or tend to the plant as required.

CONCLUSION

Thus, it is seen that the robot agent successfully performs Reinforcement Learning based path planning to accurately map the environment and its obstacles and identify the optimal path to reach a given destination. In the 3x3 environment, the robot performs accurate obstacle detection using depth estimation and, by employing the SARSA algorithm, finds the optimal path to the destination within 10 episodes.

Therefore, SARSA-based path planning has been tested using hardware implementation, and its ability to determine the optimal path to reach a destination in an unknown environment while detecting and avoiding obstacles has been validated.

FUTURE WORKS

This implementation can be extended to larger environments that are more dynamic. Future works may involve increasing the degrees of freedom of the environment and going beyond cell-based environments.

It can have diversified applications ranging from garden or warehouse management, serving food at restaurants, Air crash investigations, Search and Rescue operations and even Space Exploration.

REFERENCES

[1]   S. Zheng, and H. Liu, "Improved Multi-Agent Deep Deterministic Policy Gradient for Path Planning-Based Crowd Simulation", *IEEE Access,* vol. 7, pp. 147755-147770, 2019.
[http://dx.doi.org/10.1109/ACCESS.2019.2946659]

[2]   X. Zhou, P. Wu, H. Zhang, W. Guo, and Y. Liu, "Learn to Navigate: Cooperative Path Planning for Unmanned Surface Vehicles Using Deep Reinforcement Learning", In: *IEEE Access* vol. 7. , 2019.
[http://dx.doi.org/10.1109/ACCESS.2019.2953326]

[3]   C. Chen, J. Jiang, Lv. Ning, and S. Li, *An Intelligent Path Planning Scheme of Autonomous Vehicles Platoon Using Deep Reinforcement Learning on Network Edge, IEEE.* vol. 8. IEEE Access, 2020.
[http://dx.doi.org/10.1109/ACCESS.2020.2998015]

[4]   C. Wang, J. Wang, Y. Shen, and X. Zhang, "Autonomous Navigation of UAVs in Large-Scale Complex Environments: A Deep Reinforcement Learning Approach", *IEEE Trans. Vehicular Technol.,* vol. 68, no. 3, pp. 2124-2136, 2019.
[http://dx.doi.org/10.1109/TVT.2018.2890773]

[5]   D. Ebrahimi, S. Sharafeddine, P.H. Ho, and C. Assi, "Autonomous UAV Trajectory for Localizing Ground Objects: A Reinforcement Learning Approach", *IEEE Trans. Mobile Comput.,* vol. 20, no. 4, pp. 1312-1324, 2021.
[http://dx.doi.org/10.1109/TMC.2020.2966989]

[6]   B. Wang, Z. Liu, Q. Li, and A. Prorok, "Mobile Robot Path Planning in Dynamic Environments Through Globally Guided Reinforcement Learning", *IEEE Robot. Autom. Lett.,* vol. 5, no. 4, pp. 6932-6939, 2020.
[http://dx.doi.org/10.1109/LRA.2020.3026638]

[7]   A. Konar, I.G. Chakraborty, S.J. Singh, L.C. Jain, and A.K. Nagar, "A Deterministic Improved Q-Learning for Path Planning of a Mobile Robot", *IEEE Trans. Syst. Man Cybern. Syst.,* vol. 43, no. 5, pp. 1141-1153, 2013.
[http://dx.doi.org/10.1109/TSMCA.2012.2227719]

[8]   J. Xin, H. Zhao, D. Liu, and M. Li, "Application of deep reinforcement learning in mobile robot path planning", *Chinese Automation Congress (CAC),* 2017.
[http://dx.doi.org/10.1109/CAC.2017.8244061]

[9]   V.N. Sichkar, "Reinforcement Learning Algorithms in Global Path Planning for Mobile Robot", *International Conference on Industrial Engineering, Applications, and Manufacturing (ICIEAM),* pp. 1-5, 2019.
[http://dx.doi.org/10.1109/ICIEAM.2019.8742915]

[10]   V.N. Sichkar, "Reinforcement Learning Algorithms in Global Path Planning for Mobile Robot", *International Conference on Industrial Engineering, Applications and Manufacturing,* 2019.
[http://dx.doi.org/10.1109/ICIEAM.2019.8742915]

[11]   P. Gao, Z. Liu, Z. Wu, and D. Wang, "A Global Path Planning Algorithm for Robots Using Reinforcement Learning", *IEEE International Conference on Robotics and Biomimetics (ROBIO),* 2019.
[http://dx.doi.org/10.1109/ROBIO49542.2019.8961753]

[12] Y. Long, and H. He, "Robot path planning based on deep reinforcement learning", *IEEE Conference on Telecommunications, Optics and Computer Science (TOCS)*, 2020.
[http://dx.doi.org/10.1109/TOCS50858.2020.9339752]

[13] P. Mohan, L. Sharma, and P. Narayan, "Optimal Path Finding using Iterative SARSA", *5th International Conference on Intelligent Computing and Control Systems (ICICCS)*, pp. 811-817, 2021.

CHAPTER 14

Analysis of Human Gait by Selecting Anthropometric Data Based on Machine Learning Regression Approach

Nitesh Singh Malan[1,*] and **Mukul Kumar Gupta**[1,*]

[1] School of Engineering, University of Petroleum and Energy Studies (UPES), Dehradun, India

Abstract: This paper aims to elucidate a method to simulate human gait, which can help design a fully functional exoskeleton to rehabilitate the human lower limb. We present a method to calculate the forces and moments of each lower limb joint using human anthropometric parameters and free body diagrams. Various forces and moment of forces of lower limb joints have been calculated. The anthropometric data is evaluated using the linear regression approach. Also, in this work, we have simulated the normal human walking pattern. The forces and moments acting on lower limb joints are calculated in horizontal and vertical directions, and the human gait was simulated for a speed of 1.8m/s. The estimated results can be used as input parameters for the development of an exoskeleton for the rehabilitation of the human lower limb.

Keywords: Anthropometric data, Human Gait, Regression, Rehabilitation.

INTRODUCTION

When the movements of the lower limbs are affected by some injuries or stroke, rehabilitation treatment is required to restore gait function and regain the capacity to walk independently. A number of individual rehabilitation treatment approaches have been proposed to improve overall walking ability [1]. The most important task considered is to be re-learned, that is, the possibility to walk again [2]. In traditional techniques, manual regular treadmill training was used to improve the walking capabilities of the patients. Repetitive training requires leg movements of the patient assisted by physiotherapists, which usually limits this training [3]. A driven gait orthosis (DGO) [4] was designed to improve treadmill training for patients. DGO is a robot that provides automated locomotor training to non-ambulatory patients, which gives advantages over manual training. DGO is

* **Corresponding authors Nitesh Singh Malan and Mukul Kumar Gupta:** School of Engineering, University of Petroleum and Energy Studies (UPES), Dehradun, India; E-mails: niteshs.malan@ddn.upes.ac.in and mkgupta@ddn.upes.ac.in

Hemachandran K., Raul V. Rodriguez, Umashankar Subramaniam & Valentina Emilia Balas (Eds.)

stronger than the therapists' physical abilities and requires fewer therapists to carry out the therapy. Also, work on robotic rehabilitation has proven effective compared with manual therapy in patients suffering from a stroke. A mechanized gait trainer for gait rehabilitation was designed and constructed by Stefan Hesse and Dietmar Uhlenbrock for the repetitive practice of a gait pattern to improve the walking capabilities of stroke patients [5]. In recent developments, many wearable robotic devices have been developed to assist patients with a lack of mobility. Also, work on robotic rehabilitation has proven to be effective compared with manual therapy in patients suffering from a stroke. A mechanized gait trainer for gait rehabilitation was designed and constructed by Stefan Hesse and Dietmar Uhlenbrock for the repetitive practice of a gait pattern to improve the walking capabilities of stroke patients [5]. In recent developments, many wearable robotic devices have been developed to assist patients with a lack of mobility.

For designing any robotic device for the rehabilitation of human gait, a detailed study of human gait is required, including range of motion, forces, and moment of forces at different joints of the lower limb. A study [6] proposed a mathematical model for human movement dynamics. It argues that previous moment records can demonstrate a typical walking pattern. The combined moment of force values of the ankle, knee, and hip joints can maintain the balance of the body during the stance phase of gait. The final moment was defined as the summation of the moment of the knee joint and the difference in the moment of the hip and ankle joints [7]. A computer model was developed to calculate the force and moment of hip, knee, and ankle joint muscles [8]. The 3-d plain was used to obtain kinematics data. To maintain the stability of whole body during human gait, a model was used. It evaluated the effects of forces, acceleration and joint moments which are acting on the foot and hip, but active hip abduction moment was used to control the stability of the upper body & swing leg [9, 10]. Dynamic equations were proposed to evaluate the proximal end forces and moments with respect to the distal end forces and moments for each lower limb segment. In this method, forces acting on foot are evaluated first and continued up the limb [11]. Differential equations are used for direct dynamic modeling, making this process very complicated [12]. To interpret variations and their source and effect relationships, kinematic and kinetic patterns were studied. The hip and knee joints showed high changes in kinetic force patterns [13].

The study of human gait is broadly classified into kinematics and kinetics. Kinematics is the study of the motion of bodies with respect to time, displacement, and velocity, either in longitudinal or rotational directions. The study of the forces associated with the motion and forces resulting from the motion is known as kinetics. The study of the kinetics of human movement plays a vital role in understanding the basic mechanics of the human lower limb's movement while walking and finding the cause of deviation of any movement. The cause of deviation of any movement can be determined by estimating different patterns of the forces. The kinetics study also helps explain the method to calculate force and moment of forces using kinematic and inertial properties. The use of a free body diagram is suggested [14], in which segments are broken, and a free body diagram is used for presenting the various forces acting on the various segments of the body *(i.e.,* thigh, shank and foot). The range of motions at the hip, knee and ankle joints while people walk at normal speed should be evaluated [15]. This paper aims to simulate human gait, to estimate various parameters of lower limbs involved during walking. These estimated parameters can serve as design inputs for developing a fully functional exoskeleton for the rehabilitation of human lower limbs.

METHODS AND MATERIALS

This problem was analyzed, formulated and divided into different parts. In this work, a simple and novel way of simulating human gait is proposed. Rehabilitation of human lower limbs is possible with the help of findings gained in this work. We have proposed a method to simulate a human walking pattern in this work. First, we used the machine learning regression approach for anthropometric parameters and found the lower limb segmental length, weight, moment of inertia, *etc.,* of an average adult male. Then we utilize segmental values to estimate all the forces and moments of force acting on the lower limb joint using the differential equations. Programming has been used to calculate joint forces and moments of forces acting on the joints in horizontal and vertical directions. Using these estimated values as input parameters, we designed a 3-link model on Simulink to mimic the human walking pattern. 3 links correspond to the foot, shank and thigh, and the joints correspond to the ankle and knee joints. The estimated results can be used as input parameters for the development of an exoskeleton for the rehabilitation of the human lower limb. Newtonian mechanics method is used to study the cause of deviation during different phases of human gait. The Newton-Eular method is a mathematical approach to solve a system of equations; it starts with initial conditions and then applies input values (here, the input values are normalized segmental mass, length and acceleration) to estimate force and moment values. In this work, normalized values of mass, length, and center of mass for each lower limb segment of a normal human subject assuming weight 70Kg and height 170cm have been calculated with the help of the approximate proportion values of Dempster's table. Using the parallel axis theorem, we have also calculated the moments of inertia on the basis of the axis of rotation of different segments of lower limbs. Estimated values are tabulated in Table **1**.

Table 1. Inertial Characteristics of Lower Limb Segments Estimated for 70Kg Male.

Segment	Mass (kg)	Moment of inertia (kg.m^2) About center of gravity	Length (m)	Center of Mass (m)		Acceleration (rad/s^2)		Angular Acceleration (swing phase)
				Proximal	Distal	Ax	Ay	
Foot	1.01	0.015	0.258	0.1254	0.125	9.07	6.62	21.69
Shank	3.25	0.047	0.40	0.145	0.190	-0.03	-4.2	36.9
Thigh	7.0	0.122	0.41	0.130	0.170	1.2	-0.7	4.2

Anatomical and equivalent link segment models of lower limbs are shown Figs. (**1i & ii**), respectively. The link segment model is used to calculate the ground reaction forces and muscle moments. Whereas free body diagrams shown in Fig. (**1iii**) separate each segment from the other at the joints and the forces acting across each joint is calculated. Here an inverse dynamics technique was used to calculate the required forces and moment of forces based on the kinematics and inertial properties of the body. FBD has been used to estimate the various forces acting on the various segments of the lower limb (*i.e.,* thigh, shank, and foot).

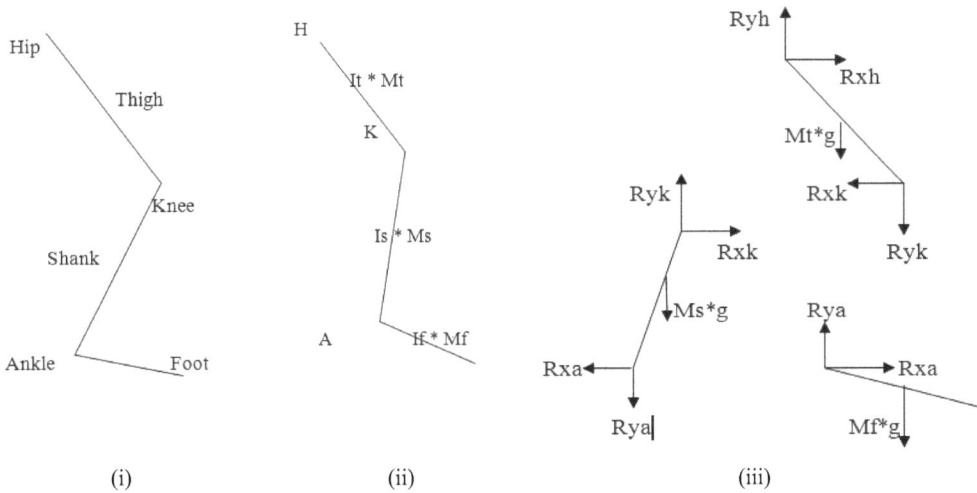

Fig. (1). (**i**) Anatomical model of human leg (**ii**) Equivalent Link segment Model and (**iii**) Free Body Diagram, broken from joints.

MULTI LINK SEGMENT MODEL

Multi segment analysis was performed to estimate the forces and moments acting on the human leg segments and joints. The analysis started at the most distal segment (foot). The foot segment has three unknown parameters (two forces acting in the x and y direction and one moment of force), and for solving these parameters, three equations were used. The thigh and shank segment has six unknown parameters each. To solve the problem, we started from the foot and calculated all three unknown parameters successfully. Then we proceeded to the next segment, the shank has six unknown parameters; out of six, there were three parameters which were already calculated from the foot segment equations. But foot forces must be used in opposite directions as per Newton's third law, which states that the forces on the distal end of one segment must be equal and opposite to those of the proximal end of the adjacent segment. In this work, Newton-Eular equations were used to calculate the effect of forces and moment of forces acting on the human lower limb joints. The anthropometric parameters as listed in Table **1** were used to calculate net reaction forces and moments. At a given point in time, a set of forces act on the body. To get the single resultant force vector, we combined the set of forces acting on the body. To estimate the single resultant force, Newton's 2nd law can be used. Here the resultant forces were divided into unknown and known forces, and the unknown forces formed a single net force which can be solved. The effects of forces and moment of forces acting at the joints were determined using the anthropometric parameters, *i.e.,* moment of inertia and mass of each segment. The calculated net forces and moment of forces illustrate the addition of the total effects of forces of respective lower limb joint segments in producing movement. During the human walking each of the lower limbs joint is responsible for the steps taken. The forces and moment of forces acting at the joints were determined by using the following equation.

SIMULATION OF HUMAN GAIT

In the following paragraphs, an approach to develop human like walking pattern has been proposed. A gait pattern has been simulated for the stance and swing phases. The human gait cycle has different phases as the cycle proceeds from 0 to 100 percent.

We have modelled the human lower limb on sim mechanics. This model includes links, joint sensors, joint initial conditions, rotational joints, and scopes. In this model, three links have been used, which are analogues to the thigh, shank and foot of the human lower limb. The length of each link is evaluated using the machine learning regression approach as discussed in the next section. These links are named as body segments in sim mechanics and also, we can alter the weight and length of each of the body segment, for instance, the estimated thigh weight as listed in Table **1** is 7.0Kg; hence we used this value for thigh body weight in the designed model [16]. Similarly, values for other body segments were applied. However, rotational joints have been used as hip, knee and ankle joints. The estimated lower limb segmental values derived in the previous section (listed in Table **1**) were applied as joint initial conditions. Here we have divided the gait cycle into different parts as a simulation of one complete gait cycle cannot be possible because the axis of action is different during different phases of the gait cycle. The three-link model has been designed on sim mechanics as shown in Fig. (**2**).

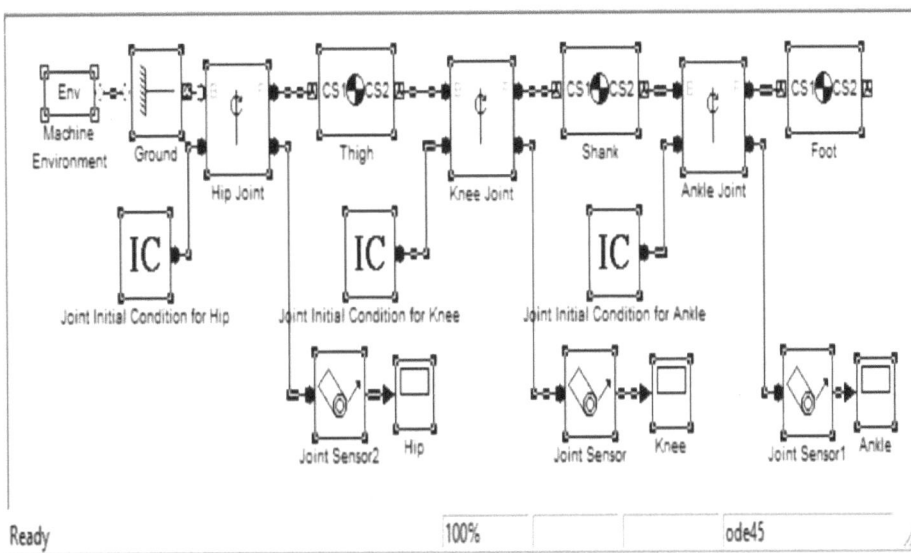

Fig. (2). Designed model of human lower limbs on sim mechanics.

REGRESSION FOR ANTHROPOMETRIC MEASUREMENTS

A linear technique for modeling the connection between a scalar response and one or more explanatory variables is known as linear regression. Simple linear regression is used when there is only one explanatory variable; multiple linear regression is used when there is more than one [17]. The lengths of the foot, shank and thigh presented in Table **1** are estimated using the machine learning regression approach. In total, 10 subjects participated in the experiment, and their length of body segments (foot, shank and thigh) was measured. The measured values are thus fitted using a multiple linear regression model [18]. For training the regression model, input variables were the length of body segments (foot, shank and thigh) and height of the participants. Using the trained model, anthropometric data is evaluated for a person with a height of 170m.

RESULTS AND DISCUSSION

Simulation has been done to calculate the various forces and moments of forces acting on the lower limb muscles and joints. The forces and moments are calculated in horizontal and vertical directions. The computed values of horizontal force, vertical force and moment are tabulated in Table **2**.

Table 2. Estimated Force and Moment Values for Lower Limbs.

Lower limb	Horizontal Joint Reaction force	Vertical Joint Reaction force (Ay)	Moment (Nm)
Joint	(AX) (N)	(N)	
Ankle	9.2	19.8840	0.9709
Knee	9.3	23.38	2.34

The moment values decrease from the ankle to the hip joint, and vertical force values increase from the distal to the proximal joint. The gait cycle was divided into different parts

1. 0 to 40% of the gait cycle
2. 40 to 50% of the gait cycle
3. 50 to 60% of the gait cycle
4. 60 to 80% of the gait cycle
5. 80 to 100% of the gait cycle

The human gait was simulated for the speed of 1.8m/s. The simulation animated results for lower limbs at 0%, 40%, 60% and 80% of the gait cycle are shown in (Fig. **3**). Also, graphs shown in Figs. (**4i & ii**) elucidate the angle of variation in degree for knee and ankle joints, respectively, during 0 to 40% of the gait cycle. In this graph, the y-axis represents the percentage of the gait and the x-axis represents the variation in the angle of motion in degree. Graphs for the rest of the gait cycle were also obtained. Table **3** lists the results obtained from the graphs, which show the angle of the knee and ankle at different instants of the gait cycle. All the angles were measured in the sagittal plane. For validation, these graphs were compared with the sagittal plane angles of normal adults during walking. The ranges of angles obtained after simulations were found very close to the lower limb joint angle ranges during real-time gait.

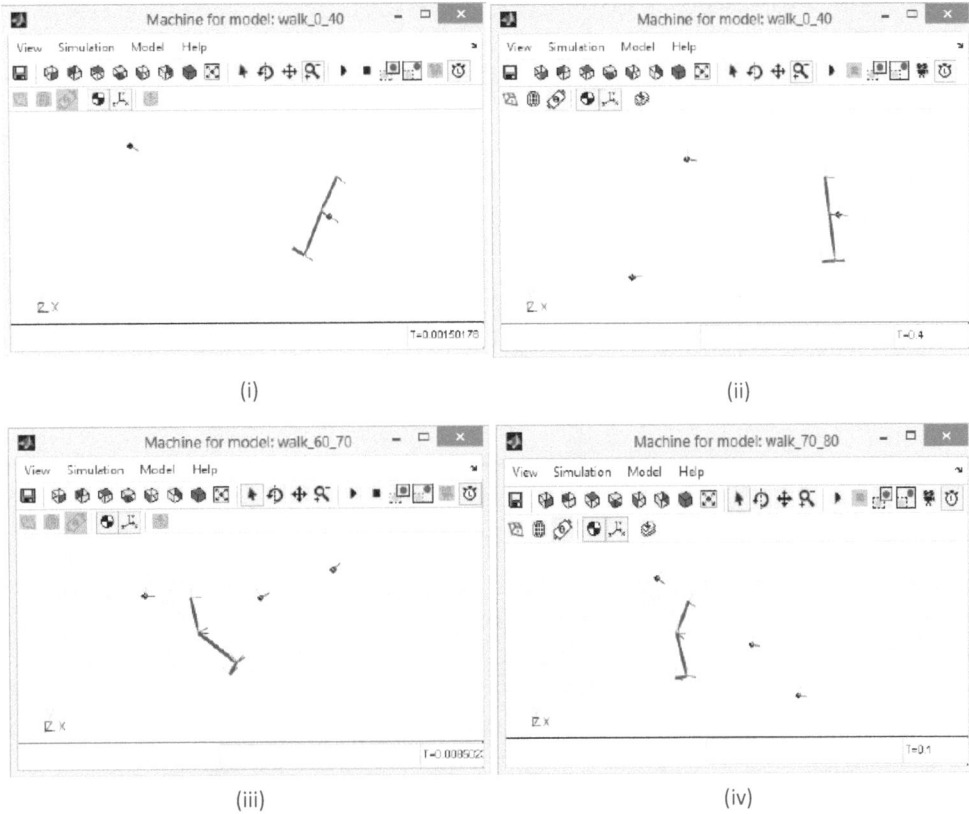

Fig. (3). The simulation animated results for lower limbs joint angles during walking at (**i**) 0%, (**ii**) 40%, (**iii**) 60% and (**iv**) 80% respectively, of the gait cycle.

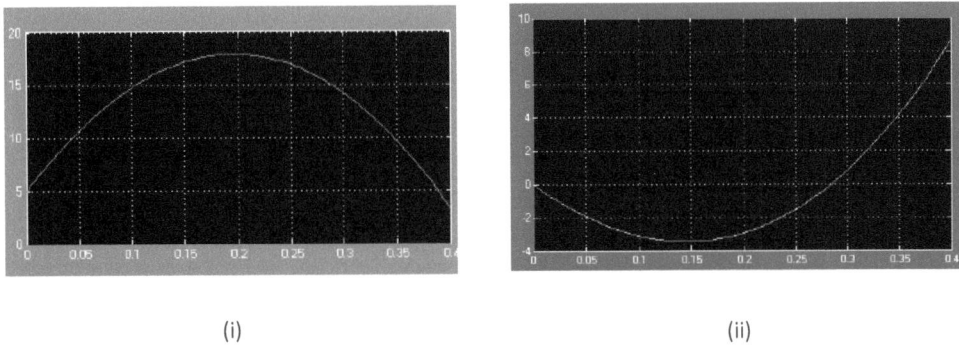

Fig. (4). (**i**) Variations in Knee joint sagittal plane angles for 0 to 40% of the gait cycle. (**ii**) Variations in Ankle joint angle for 0 to 40% of the gait cycle.

Table 3. Estimated values indicating Variation in Knee and Ankle.

Phase of Gait cycle (in %)	Knee angle (in degree)	Ankle angle(in degree)
0	5	0
20	18	-3
40	3	8
60	32	-13
70	52	-8.5
80	28	1
90	17	23
100	2	1

CONCLUSION

In this project, a method to calculate the force and moment of force of each lower limb joint of a human has been presented using human anthropometric parameters and free body diagrams. This is performed to give an idea of the limitation of force bearing values of lower limb prosthetics. The developed Matlab code can effectively calculate various forces and moments of forces acting on lower limb joints during gait. To simulate walking patterns for one complete gait cycle, we have divided the different phases of gait and simulated each phase individually. The estimated results can be used as input parameters for the development of an exoskeleton for the rehabilitation of the human lower limb. Exoskeletons are wearable devices to assist patients with neurological conditions. Therapists use exoskeletons to assist persons with spinal cord injury that affects limb movement. The range of angle movements for lower limb joints obtained in this work is found to be very close to the real-time walking movements at a constant speed. An intelligent prosthetic leg using a microcontroller and actuators for knee movement rehabilitation can be designed.

ACKNOWLEDGEMENTS

The authors would like to express their heartfelt gratitude to the Electrical cluster, School of Engineering, UPES Dehradun, for providing valuable resources to conduct this work.

REFERENCES

[1] S. Holland, S.D. Silberstein, F. Freitag, D.W. Dodick, C. Argoff, and E. Ashman, "Evidence-based guideline update: NSAIDs and other complementary treatments for episodic migraine prevention in adults: report of the Quality Standards Subcommittee of the American Academy of Neurology and the American Headache Society", *Neurology,* vol. 78, no. 17, pp. 1346-1353, 2012.
[http://dx.doi.org/10.1212/WNL.0b013e3182535d0c] [PMID: 22529203]

[2] J.C. Furlan, V. Noonan, A. Singh, and M.G. Fehlings, "Assessment of disability in patients with acute traumatic spinal cord injury: a systematic review of the literature", *J. Neurotrauma,* vol. 28, no. 8, pp. 1413-1430, 2011.
[http://dx.doi.org/10.1089/neu.2009.1148] [PMID: 20367251]

[3] R. Riener, L. Lünenburger, S. Jezernik, M. Anderschitz, G. Colombo, and V. Dietz, "Patient-cooperative strategies for robot-aided treadmill training: first experimental results", *IEEE Trans. Neural Syst. Rehabil. Eng.,* vol. 13, no. 3, pp. 380-394, 2005.
[http://dx.doi.org/10.1109/TNSRE.2005.848628] [PMID: 16200761]

[4] G. Colombo, M. Joerg, R. Schreier, and V. Dietz, "Treadmill training of paraplegic patients using a robotic orthosis", *J. Rehabil. Res. Dev.,* vol. 37, no. 6, pp. 693-700, 2000.
[PMID: 11321005]

[5] S. Hesse, and D. Uhlenbrock, "A mechanized gait trainer for restoration of gait", *J. Rehabil. Res. Dev.,* vol. 37, no. 6, pp. 701-708, 2000.
[PMID: 11321006]

[6] S. Onyshko, and D.A. Winter, "A mathematical model for the dynamics of human locomotion", *J. Biomech.,* vol. 13, no. 4, pp. 361-368, 1980.
[http://dx.doi.org/10.1016/0021-9290(80)90016-0] [PMID: 7400165]

[7] D.A. Winter, "Overall principle of lower limb support during stance phase of gait", *J. Biomech.,* vol. 13, no. 11, pp. 923-927, 1980.
[http://dx.doi.org/10.1016/0021-9290(80)90162-1] [PMID: 7275999]

[8] E.M. Arnold, S.R. Ward, R.L. Lieber, and S.L. Delp, "A model of the lower limb for analysis of human movement", *Ann. Biomed. Eng.,* vol. 38, no. 2, pp. 269-279, 2010.
[http://dx.doi.org/10.1007/s10439-009-9852-5] [PMID: 19957039]

[9] G. Bharatkumar, K. Daigle, M. Pandey, Q. Cai, and J. Aggarwal, "Lower limb kinematics of human walking with the medial axis transformation", *Am J Phys Med,* vol. 46, no. 1, pp. 290-297, 1996.
[http://dx.doi.org/10.1109/MNRAO.1994.346252]

[10] C.D. MacKinnon, and D.A. Winter, "Control of whole body balance in the frontal plane during human walking", *J. Biomech.,* vol. 26, no. 6, pp. 633-644, 1993.
[http://dx.doi.org/10.1016/0021-9290(93)90027-C] [PMID: 8514809]

[11] B. Bresler, and J.P. Frankel, "The Forces and Moments in the Leg During Level Walking", *Trans. Am. Soc. Mech. Eng.,* vol. 72, no. 1, pp. 27-36, 1950.
[http://dx.doi.org/10.1115/1.4016578]

[12] R. Crowninshield, M.H. Pope, and R.J. Johnson, "An analytical model of the knee", *J. Biomech.,* vol. 9, no. 6, pp. 397-405, 1976.
[http://dx.doi.org/10.1016/0021-9290(76)90117-2] [PMID: 932053]

[13] M.L. Audu, and D.T. Davy, "The influence of muscle model complexity in musculoskeletal motion modeling", *J. Biomech. Eng.,* vol. 107, no. 2, pp. 147-157, 1985.
[http://dx.doi.org/10.1115/1.3138535] [PMID: 3999711]

[14] C.S. Putcha, and J.A. Hodgdon, "Derivation of Differential Equations of Motion for Three-segment and Foursegment Human Locomotive Models using Free-body Diagrams", *In Proceedings of the 22nd IASTED International Conference on Modelling, Identification, and Control,* pp. 523-526, 2003.

[15]   M.P. Kadaba, H.K. Ramakrishnan, and M.E. Wootten, "Measurement of lower extremity kinematics during level walking", *J. Orthop. Res.*, vol. 8, no. 3, pp. 383-392, 1990.
[http://dx.doi.org/10.1002/jor.1100080310] [PMID: 2324857]

[16]   N. Singh Malan, and S. Sharma, "Time window and frequency band optimization using regularized neighbourhood component analysis for Multi-View Motor Imagery EEG classification", *Biomed. Signal Process. Control,* vol. 67, p. 102550, 2021.
[http://dx.doi.org/10.1016/j.bspc.2021.102550]

[17]   N.S. Malan, and S. Sharma, "Feature selection using regularized neighbourhood component analysis to enhance the classification performance of motor imagery signals", *Comput. Biol. Med.,* vol. 107, pp. 118-126, 2019.
[http://dx.doi.org/10.1016/j.compbiomed.2019.02.009] [PMID: 30802693]

[18]   D.C. Montgomery, E.A. Peck, and G.G. Vining, *Introduction to linear regression analysis.* John Wiley & Sons, 2021.

SUBJECT INDEX

T

Techniques 51, 79, 81, 84, 92, 95, 100, 115,
    116, 117, 151, 152, 189, 198, 212
  contemporary 189
  conventional visualization 81
  data processing 117
  deep neural network implementation 51
  inverse dynamics 212
Technology 58, 59, 81
  image recognition 59
  powerful computer 81
  voice processing 58
Textile businesses 135
Textile industries 106, 112, 113, 118, 119,
    133, 135
  domestic 133
Thermodynamics 20
Traditional reinforcement learning techniques
    199
Transportation, smart 150
Treadmill training 209
Tree-seed algorithm (TSA) 152
Tuberculosis 52
Turing machine 18

U

UV radiation 141

V

Values, thalassemia 192
Vendor-neutral process 49
Video 33, 89, 93, 95, 96, 99
  cameras 33
  frames 89, 93, 95, 96, 99
  scene 96
Violence detection 94, 104
  method 94
  model, real-time 104
Virtual canvas 15
Virus-resistant methods, developing 52

W

Weapons, biological 11

www.ingramcontent.com/pod-product-compliance
Lightning Source LLC
Chambersburg PA
CBHW050831220326
41598CB00006B/349